U0123757

北京常见鸟类100种

100 Common Birds in Beijing

主　编：关翔宇　曹吉鑫　杨欣宇

副主编：李进宇　韩思雨　苑　超
张钰舒　马俊丽　关雪燕

Editors-in-chief
Guan Xiangyu　Cao Jixin　Yang Xinyu

Deputy Editors-in-chief
Li Jinyu　Han Siyu　Yuan Chao
Zhang Yushu　Ma Junli　Guan Xueyan

中国林業出版社
·北　京·

China Forestry Publishing House
Beijing

图书在版编目（CIP）数据

北京常见鸟类100种 / 关翔宇, 曹吉鑫, 杨欣宇主编; 李进宇等副主编. -- 北京 : 中国林业出版社, 2024.1
ISBN 978-7-5219-2485-5

Ⅰ. ①北… Ⅱ. ①关… ②曹… ③杨… ④李… Ⅲ. ①鸟类－北京－图谱 Ⅳ. ①Q959.708-64

中国国家版本馆CIP数据核字（2024）第003691号

审图号：京S（2024）001号

责任编辑：刘香瑞
装帧设计：柴鉴云

出版发行　中国林业出版社
（100009，北京市西城区刘海胡同7号，电话 010-83143545）
网　　址　www.forestry.gov.cn/lycb.thml
印　　刷　北京雅昌艺术印刷有限公司
版　　次　2024年1月第1版
印　　次　2024年1月第1次
开　　本　880mm × 1230mm　1/32
印　　张　8.75
字　　数　350千字
定　　价　98.00元

编委会

主　编

关翔宇　曹吉鑫　杨欣宇

副主编

李进宇　韩思雨　苑　超　张钰舒　马俊丽　关雪燕

编　委

周文仪　张逐阳　文　辉　张　琴　柴鉴云　郑秋旸　史　洋　楚建梅
叶元兴　张晓川　张友敬　李新源　周欣楠　张　姝　陈　晓　曹　海
吴建芝　王　欢　闫　岩　高俊虹　韩　艺　张京州　顿媛媛　戴日新

摄　影

文　辉　关翔宇　赵云天　沈　岩　娄方洲　万　伟　张　龙　罗义华

Contributors

Editors-in-chief

Guan Xiangyu　Cao Jixin　Yang Xinyu

Deputy Editors-in-chief

Li Jinyu　Han Siyu　Yuan Chao　Zhang Yushu　Ma Junli　Guan Xueyan

Editorial Members

Zhou Wenyi　Zhang Zhuyang　Wen Hui　Zhang Shen　Chai Jianyun
Zheng Qiuyang　Shi Yang　Chu Jianmei　Ye Yuanxing　Zhang Xiaochuan
Zhang Youjing　Li Xinyuan　Zhou Xinnan　Zhang Shu　Chen Xiao
Cao Hai　Wu Jianzhi　Wang Huan　Yan Yan　Gao Junhong　Han Yi
Zhang Jingzhou　Dun Yuanyuan　Dai Rixin

Photographers

Wen Hui　Guan Xiangyu　Zhao Yuntian　Shen Yan　Lou Fangzhou
Wan Wei　Zhang Long　Luo Yihua

序一

1996 年，北京地区的民间环保组织“自然之友”和“绿家园”开始组织民间观鸟活动。它们是我国内地最早有组织地开展观鸟普及的先行力量。当时人们不了解也不理解什么是观鸟。鸟有什么好看的？如果真想看鸟，就去动物园或是养鸟者那去看不就得了？

经过 20 多年的普及，社会及广大市民已对观鸟有所了解，民间观鸟组织和观鸟活动遍地开花。有些地区，观鸟不仅仅是停留在个人的休闲娱乐活动层面，而是已成为时尚并被鸟友们不断地推陈出新。观鸟者们分享着观鸟的经验与心得。在条件成熟的地区，专业人士或资深观鸟人组织有能力的鸟友，还能为当地开展鸟类的普查和动态监测。观鸟者们身体力行地推动着公民科学，成为一支践行生态文明建设的强壮队伍。

现代观鸟活动已经持续了 200 多年，它起源于英国及欧洲的一些其他国家，后传播到世界各地。观鸟者中的绝大多数人都是非专业人员，他们从事各种职业，但一旦学会观鸟就会将其作为一种爱好。我们知道，每个人的一生都会有许多爱好，如果能把观鸟作为一生的爱好，一定会使你受益匪浅。

观鸟的过程，就像是在读一本读不完的自然故事书，每当我们在野外见到熟识的鸟，那如同是老友重逢，若见到未曾见到过的鸟，那就是又结识了新朋友。观鸟不只是能叫出鸟的名字，而是要去解读鸟和其他物种的关系，探究它们的自然历史，还要弄懂不同鸟种赖以生存的独特环境，以及它们在自然舞台上表现出来的各种行为。也就是说，观鸟要在不同的维度上观，要学会在观鸟的过程中，观形态，观行为，观生态。

国外观鸟界流行着这样一句话："学会观鸟，就如同获得了一张通往自然剧院的终身免费门票。"我们是社会的人，也是自然的人。在当今高速发展的时代，人们注重科技的发展和经济的振兴，很容易脱离自然的体验。而观鸟却能帮助人们在繁忙的工作之余，投入大自然中，丢掉烦恼、愉悦心情、修养身心、认识世界、解读生命。

北京是世界上鸟类最丰富的首都城市之一，已经记录到 500 多种野生鸟类，它们和我们共同生活在一起，我们若稍加注意，就会发现这些鸟类的生活是非常有趣的，并且这些鸟类和我们有着各种各样的联系。我们有责任去关心和保护它们的生存环境。

不管你是常驻北京的居民，还是来北京长期或短暂停留的朋友，都有机会到城市公园或北京郊野去欣赏那些美丽而灵动的鸟类。

无论你是六七岁的孩童，还是六七十岁的老人，走进鸟类世界，学习观鸟都是非常适宜的活动。从小学生到退休老人，观鸟不会受到年龄的限制。

让我们共同努力，去与鸟类为伴，成为热爱自然和敬畏生命的文明使者。

赵欣如

2023 年 8 月 2 日

于北京师范大学

Foreword 1

In 1996, non-governmental organizations in Beijing such as "Friends of Nature" and "Green Home" started to organize private bird-watching activities. They were among the pioneering forces to systematically promote bird-watching in Chinese mainland. At that time, people lacked understanding and appreciation for bird-watching. What's so fascinating about bird? If you really want to see birds, why not go to the zoo or visit pet markets?

After more than 20 years of popularization, society and the general public have gained some understanding of bird-watching, and non-governmental bird-watching organizations and activities have sprung up everywhere. In some regions, bird-watching has transcended individual leisure and entertainment, becoming a fashion trend continuously innovated by birdwatchers. birdwatchers share their experiences and insights. In areas where conditions are favorable, professionals or experienced birdwatchers organize capable birders to conduct bird surveys and dynamic monitoring. birdwatchers actively promote citizen science, becoming a robust force in practicing ecological civilization construction.

Modern bird-watching has been ongoing for over 200 years, originating in the United Kingdom and other European countries before spreading worldwide. The majority of birdwatchers are non-professionals engaged in various occupations, but once they have learnt the art of bird-watching, it becomes a lifelong hobby. We know that everyone has had numerous hobbies throughout their lives, and if bird-watching can be embraced as a lifelong passion, it is sure to bring immeasurable benefits.

The process of bird-watching is akin to reading an endless book of nature stories. Each time we encounter familiar birds in the wild, it feels like a reunion with old friends. If we come across birds we have never seen before, it is like making new acquaintances. Bird-watching is not just about being able to name the birds, it involves interpreting the relationships between birds and other species, delving into their natural history. It also requires understanding the unique environments that different bird species depend on for survival, as well as the various behaviors they exhibit on the natural stage. In other words, bird-watching requires observing in different dimensions, learning to observe the morphology, behavior, and ecology of birds during the process.

In the world of bird-watching abroad, there is a popular saying: " Learning to bird is like getting a lifetime ticket to the theater of nature". We are social beings, but we are also part of nature. In today's rapidly advancing era, where emphasis is placed on technological development and economic revitalization, it is easy for people to become disconnected from nature. However, bird-watching, can assist individuals take a break from daily hustle and bustle, immerse themselves in nature, shed their worries, enhance their mood, cultivate physical and mental well-being, gain insights into the world, and decipher the mysteries of life.

Beijing is one of the world's capitals with the richest bird diversity, with over 500 recorded species of wild birds. They coexist with us. And if we pay a little attention, we can discover the fascinating lives of these birds and the various connections they share with us. It is our responsibility to care for and protect their living environment.

Whether you are a resident or a visitor of Beijing, there is an opportunity for you to appreciate the beautiful and agile birds in urban parks or the outskirts of this fascinating metropolis.

Whether you are in six or seven-year-old child or a person in your sixties or seventies, entering the world of birds and learning bird-watching is a highly suitable activity. From primary school students to retired elders, bird-watching is not constrained by age.

Let's make a collective effort to accompany birds, becoming ambassadors of civilization who awe nature, cherish life, and uphold social values.

Zhao Xinru
August 2, 2023
Beijing Normal University

序二

北京是中国的首都，也是鸟类多样性非常丰富的大都市之一。根据北京市园林绿化局发布的《北京市陆生野生动物名录(2023)》，北京市现有野生鸟类515种，大约占全国鸟类物种总数的三分之一。在二十国集团国家的首都中，北京的鸟类物种数量排名第二，仅次于巴西的首都巴西利亚。

鸟类大多白天活动，因其广布性、环境敏感性和易观测等特征，成为反映城市生物多样性的重要指示类群。鸟的种类、数量以及行为活动规律上的变化都可以作为评价当地环境质量的依据。北京地处东亚－澳大利西亚的候鸟迁飞通道，是很多候鸟在春、秋两季迁徙时所利用的重要栖息地，也是某些珍稀候鸟繁殖或越冬的重要区域。

进入二十一世纪以来，北京市持续开展了造林绿化、湿地保护、生态修复等保护行动，吸引了众多珍稀鸟类来此栖息和繁衍，丰富了北京的鸟类多样性。例如，世界极危物种、国家一级重点保护野生动物黄胸鹀在奥林匹克森林公园、圆明园遗址公园和国家植物园等地频繁出现；国家一级重点保护野生动物黑鹳从几年前的五六十只发展到现在的100只左右；大天鹅、鸳鸯、苍鹰等珍稀鸟类逐步成为北京的“常客”，种群数量显著增多。随着北京市生态环境持续向好，市民对鸟类的保护意识也在明显提升，爱鸟护鸟在京华大地上蔚然成风。

《北京常见鸟类100种》精选了北京地区最常见、最受市民关注或极具重要保护价值的100种鸟类，详细介绍了每个鸟种的形态特征、行为习性、分布区域和栖息生境，并附有精美的图片。该名录不仅可以服务于北京地区的自然保护、园林绿化和鸟类监测站等机构，还能够服务于广大市民、鸟类爱好者，作为其学习鸟类科普知识的工具书。

加强鸟类保护，呵护生态家园，是推进生态文明建设的必然要求。我衷心希望《北京常见鸟类 100 种》的出版，能够吸引更多公众尤其是青少年参与到观鸟活动、科学调查和爱鸟护鸟行动中来，大家一起来推动北京的生物多样性保护工作再上新台阶。

张正旺

北京师范大学教授

中国动物学会副理事长

2023 年 8 月 7 日

Foreword 2

Beijing is the capital of China and one of the metropolises with rich bird diversity. According to the *List of Terrestrial Wildlife in Beijing (2023)* issued by the Beijing Municipal Forestry and Parks Bureau (Office of Beijing Greening Commission), there are 515 species of wild birds in Beijing, accounting for about one-third of the total number of bird species in China. Among the capitals of the G20 member countries, Beijing ranks second in the number of bird species, after Brasilia, the capital of Brazil.

Birds, which are mostly active during the daytime, have become an important indicator of urban biodiversity because they are widely distributed, environmentally sensitive and easy to observe. Changes in bird species, numbers and behavioral patterns can be used as bases for evaluating local environmental quality. Beijing, located on the East Asian-Australasian flyway, is an important habitat for many migratory birds during their spring and fall migrations, as well as an important area for certain rare migratory birds to breed or overwinter.

Since the 21st century, Beijing has continued to carry out afforestation, wetland protection, ecological restoration and other protective actions, attracting many rare birds to live and breed here, enriching the bird diversity of Beijing. For example, the *Emberiza aureola*, a world critically endangered species and a wild animal under China's first-class state priority conservation, appears frequently in the Olympic Forest Park, Yuanmingyuan Ruins Park, China National Botanical Garden, etc.; The *Ciconia nigra*, a national category I key protected wildlife, has grown from 50 to 60 a few years ago to around 100 now; Rare birds such as *Cygnus cygnus*, *Aix galericulata* and *Accipiter gentilis* have gradually become "regular visitors" to Beijing, with a significant increase in their population. As the ecological environment in Beijing continues to improve, the public's awareness of bird protection is also rising significantly, and the trend of loving and protecting birds has become popular in Beijing.

100 Common Birds in Beijing features 100 bird species that are the most common, the most concerned by the public, or of great conservation value in Beijing, and introduces in detail the morphological characteristics, behavioral habits, distribution areas and habitat of each bird species, with beautiful pictures. This book can not only serve institutions such as nature conservation agencies, forestry and parks bureaus and bird monitoring stations in Beijing, but can also serve the general public and bird enthusiasts as a tool for them to learn popular science knowledge about birds.

Enhancing bird conservation measures and protecting the ecological environment of our homeland are inevitable requirements for promoting the construction of ecological civilization. I sincerely hope that the publication of *100 Common Birds in Beijing* will attract more public, especially adolescents, to participate in bird-watching activities, scientific investigations and actions of loving and protecting birds. Together, we can promote the biodiversity conservation works in Beijing to a new level!

Zhang Zhengwang
Professor, Beijing Normal University
Vice Chairman, China Zoological Society
August 7, 2023

前言

北京，是中华人民共和国的首都，也是世界著名古都和现代化国际城市。北京地处华北大平原的北部，东面与天津市毗连，其余均与河北省相邻，总面积 16410.54 平方公里，全市共辖 16 个区，地势西北高、东南低。北京市西部为西山（属太行山脉），北部和东北部为军都山（属燕山山脉），两山在昌平南口相交，形成一个向东南展开的半圆形大山弯，人们称之为“北京湾”。北京市的主要河流有永定河、潮白河、北运河、拒马河等。北京市的气候为暖温带半湿润半干旱季风气候，夏季高温多雨，冬季寒冷干燥，春、秋短促。全市森林覆盖率 44.8%，森林蓄积量 31.64×10^6 立方米，城市绿化覆盖率 49.3%，人均公园绿地面积 16.63 平方米。

北京市位于全球九大候鸟迁徙通道之一的东亚－澳大利西亚通道上，是灰鹤、大鸨、青头潜鸭、中华秋沙鸭等大量珍稀濒危鸟类的重要越冬地与迁徙驿站。作为大量候鸟的停歇、繁殖和越冬地，北京对于迁徙鸟类的保护以及生物多样性的维持具有重要意义。近年来，北京持续开展系列生态系统修复工作，提升城市生物多样性，提高山区森林质量，建设大尺度生态板块，恢复湿地生态系统，连通碎片化的“生态孤岛”，打通野生动物迁移通道。截至 2023 年 4 月 15 日，《北京市陆生野生动物名录（2023）》收录的野生动物种类已经由 596 种增加至 608 种，新增加的 12 种全部为鸟类。北京的总面积仅占中国国土面积的 0.175%，而北京的鸟类种数占全国的 34.8%，很重要的原因是北京近些年的生态保护修复工作取得了显著成效。黑鹳、大天鹅、北京雨燕等在北京市的分布地不断扩大，有“鸟中大熊猫”之称的震旦鸦雀在房山、大兴、丰台等地频现，而消失近 80 年的栗斑腹鹀也再现密云水库。

本书依据综合遇见率和被记录频率选取了北京地区常见鸟类共计100种，分属18目41科76属，按照陆禽、游禽、涉禽、攀禽、猛禽、鸣禽六大生态类群依次呈现。每个鸟种精选2~3幅彩色照片，反映其形态特征和栖息生境，并以中英文对照的形式，简要记述鸟类的形态特征、分布与生境、生态习性、居留类型、保护级别等，方便使用者快速查阅。

特别感谢亚洲基础设施投资银行对本书出版的资助。

由于编著者水平有限，错漏之处实为难免，欢迎读者批评指正。

编著者

2023年8月15日

Preface

Beijing, the capital of the People's Republic of China, is also a world-famous ancient capital and a modern international city. Beijing is located in the northern part of the North China Plain, bordering Tianjin to the east and Hebei Province to the rest, with a total area of 16,410.54 km^2 and 16 districts under its jurisdiction. The terrain of Beijing is high in the northwest and low in the southeast. In the west of Beijing are the Xishan mountains, which are part of the Taihang Mountains, and in the north and northeast are the Jundu Mountains, which are part of the Yanshan Mountains. The two mountains intersect at the Nankou of Changping District, forming a semicircle bend to the southeast, which is called the "Bay of Beijing". The main rivers in Beijing include the Yongding River, Chaobai River, North Canal, Juma River, etc. The climate of Beijing is warm temperate semi-humid and semi-arid monsoon, with high temperatures and rainy summers, cold and dry winters, and short springs and autumns. The city's forest coverage is 44.8%, forest storage volume is 31.64×10^6 m^3, urban greening coverage rate is 49.3%, and per capita park green space area is 16.63 m^2.

Beijing, located on the East Asian-Australasian flyway, one of the nine major flyways for migratory birds in the world, is an important wintering place and migration station for a large number of rare and endangered birds, such as *Grus grus*, *Otis tarda*, *Aythya baeri*, *Mergus squamatus*, etc. As a resting, breeding and wintering site for a large number of migratory birds, Beijing plays a crucial role in the conservation of migratory birds and the maintenance of biodiversity. In recent years, Beijing has continued to carry out a series of ecosystem restoration efforts to enhance urban biodiversity, improve the quality of mountainous forests, build large-scale ecological panels, restore wetland ecosystems, connect fragmented "ecological islands", and open up migration corridors for wildlife. As of 15 April 2023, the number of wildlife species included in the

List of Terrestrial Wildlife in Beijing (2023) has increased from 596 to 608, and the 12 newly added species are all birds. Beijing's total area accounts for only 0.175% of China's land area, while it has 34.8% of the country's bird species. An important reason for this is that Beijing's ecological protection and restoration work in recent years has made remarkable achievements. *Ciconia nigra*, *Cygnus cygnus*, and *Apus apus pekinensis* have been expanding their distribution in the city, the *Paradoxornis heudei*, also known as the "giant panda among birds", has appeared frequently in Fangshan, Daxing, and Fengtai District, etc., and the *Emberiza jankowskii*, which has disappeared for nearly 80 years, has also reappeared in the Miyun Reservoir.

Based on the combined encounter rate and recording frequency, the present book contains 100 species of common birds in Beijing, which belong to 18 orders, 41 families and 76 genera. This book is presented in categories of six functional groups: landfowl, waterfowl, wading birds, scansorial birds, raptors and passerines. Each species is illustrated with two or three colorful photographs reflecting its morphological features and habitat, supplemented by a brief description of its morphological characteristics, distribution and habitat, behavior and ecology, type of residence, and level of protection in both Chinese and English for quick reference by users.

Special thanks to the Asian Infrastructure Investment Bank for funding this book.

Due to the limited level of the editors, if there are any inevitable errors and omissions, we invite readers to criticize and correct us.

Authors

August 15, 2023

本书使用说明

本书使用的鸟类分类系统及中文、英文和学名主要依据《中国鸟类分类与分布名录》第四版（郑光美，2023）。各鸟种的保护级别，北京市级的按照 2022 年 12 月 31 号北京市园林绿化局、北京市农业农村局发布的《北京市重点保护野生动物名录》标注；国家级的按照国家林业和草原局、农业农村部 2021 年 2 月发布的《国家重点保护野生动物名录》标注；同时标注了各种鸟在国际自然保护联盟（IUCN）红色名录的濒危级别。

读者可以通过目录查找到每一个鸟种所在的页码。鸟种页面描述与呈现的信息包括：鸟种名称、识别特征、分布与生境、生态习性、保护级别和鸟种照片，鸟种照片上标注了识别要点。

Instruction for This Book

The naming (scientific names, common names in both Chinese and English) and taxonomy of birds in this book mainly follow *A Checklist on the Classification and Distribution of the birds of China: Fourth Edition* (Zheng Guangmei, 2023). Municipal protected species are noted in accordance with third announcement of 2022, *List of Wild Animals under Beijing Priority Conservation*, issued by the Beijing Municipal Forestry and Parks Bureau and Beijing Municipal Bureau of Agriculture and Rural Affairs in December 31, 2022. National protected species are noted in accordance with *China's List of Wild Animals under state Priority Conservation* by the National Forestry and Grassland Administration and the Ministry of Agriculture and Rural Affairs in February 2021. The conservation status of each bird species according to the International Union for Conservation of Nature (IUCN) Red List of Threatened Species is also noted.

Audiences can find the page number of each bird species account through the index. Each species account includes species name, morphological traits, distribution and habitat, behaviour and ecology, conservation status and photographs. Key characteristics of certain species are listed on their photos.

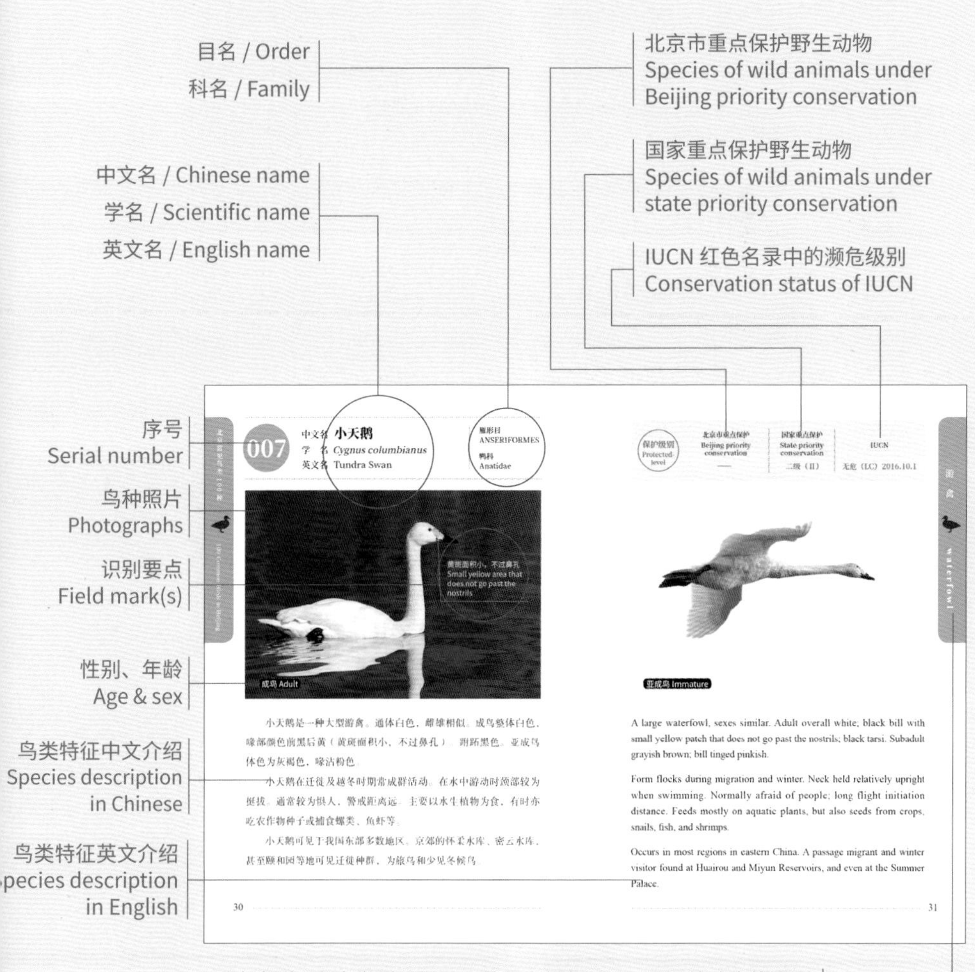

鸟类身体部位示意图
Bird Topography

全身羽区 / Topography

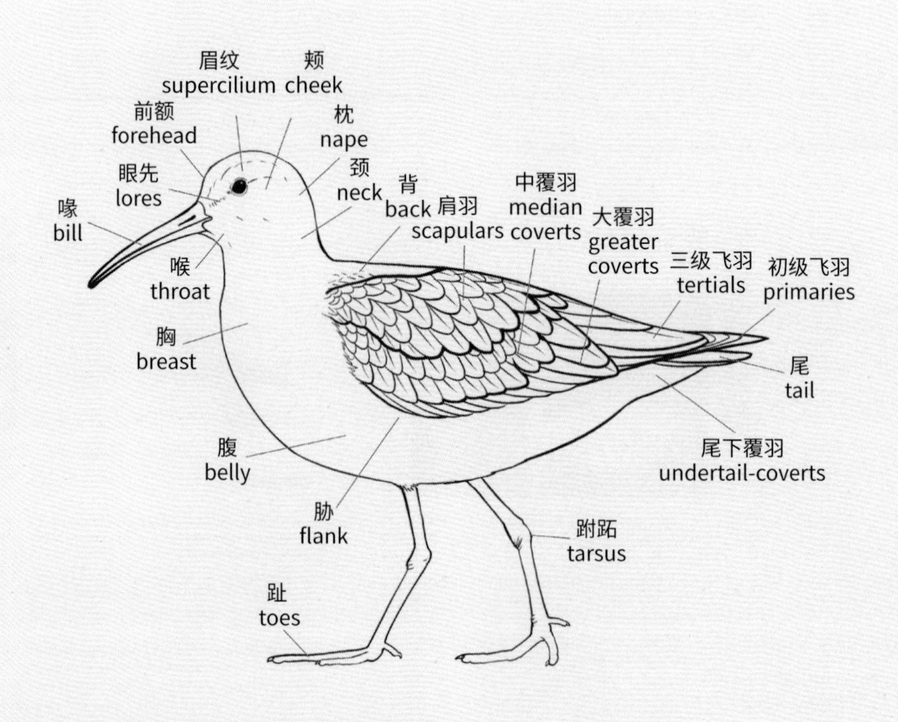

翅 / Wing

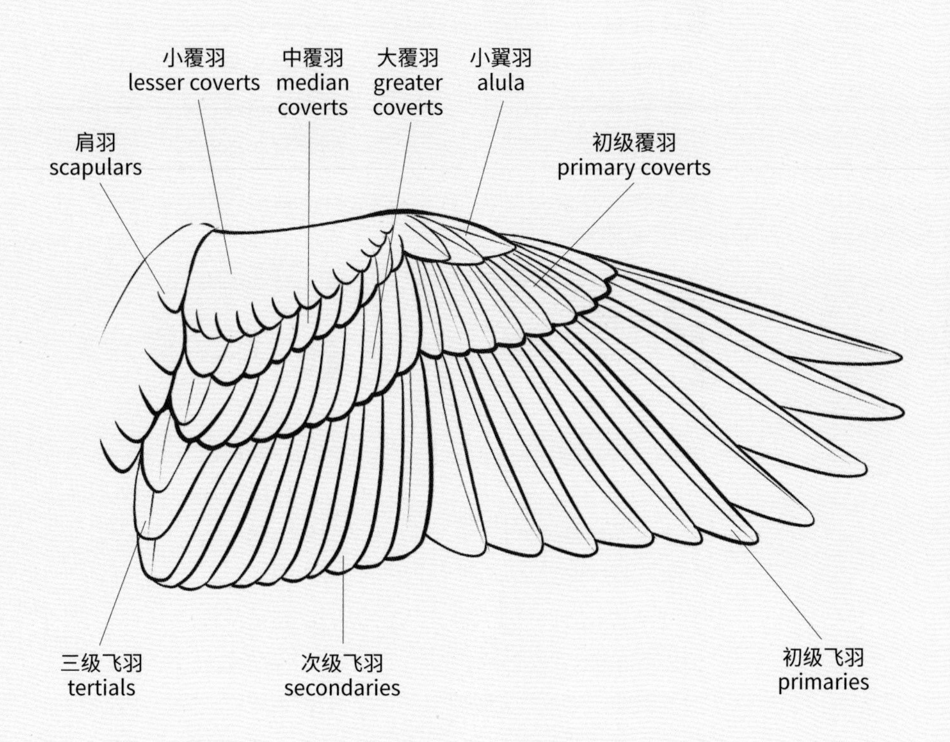
小覆羽
lesser coverts
中覆羽
median coverts
大覆羽
greater coverts
小翼羽
alula
肩羽
scapulars
初级覆羽
primary coverts
三级飞羽
tertials
次级飞羽
secondaries
初级飞羽
primaries

目录

Contents

Scansorial Birds

Raptors

Passerines

北京的自然环境

一、北京自然地理概况

北京位于华北平原的西北端，太行山余脉的西山和燕山山脉的军都山在南口关沟相交，形成一个向东南展开的半圆形大山弯，它所围绕的小平原东南部为缓慢向渤海延伸的扇形平原，又称“北京湾”。北京与天津相邻，并与天津一起被河北省环绕。北京的地理坐标为北纬 39° 38′ ~41° 05′、东经 115° 25′ ~117° 30′；总面积 16410.54 平方公里；最高峰为门头沟区的东灵山，海拔 2303 米。2020 年，北京行政区划调整为 16 个区：东城区、西城区、朝阳区、丰台区、石景山区、海淀区、门头沟区、房山区、通州区、顺义区、昌平区、大兴区、怀柔区、平谷区、密云区、延庆区。

北京的气候为暖温带半湿润半干旱季风气候，夏季高温多雨，冬季寒冷干燥，春、秋短促。年平均气温在 8~12 摄氏度。年均降水量 600 毫米，降水季节分配很不均匀，全年降水的 70%~80% 集中在夏季 6~8 月。

二、北京鸟类的生态类群及分布

北京的鸟类包括陆禽、游禽、涉禽、攀禽、猛禽、鸣禽等六大生态类群。在山区和平原，以及城市中心的各种生境中，各生态类群都有相同或不同代表种类的分布。由于鸟类善于飞行，它们选择栖息地的能力很强，分布往往随季节、食物等因素的变化而变化。

1. 陆禽

陆禽的后肢强壮，适于地面行走；翅短圆；喙强壮且多为弓形，适于啄食。代表种类有环颈雉 (*Phasianus colchicus*)、鹌鹑 (*Coturnix japonica*) 等。斑鸠和鸽虽然善飞翔，但取食主要在地面，因此也被归于陆禽。

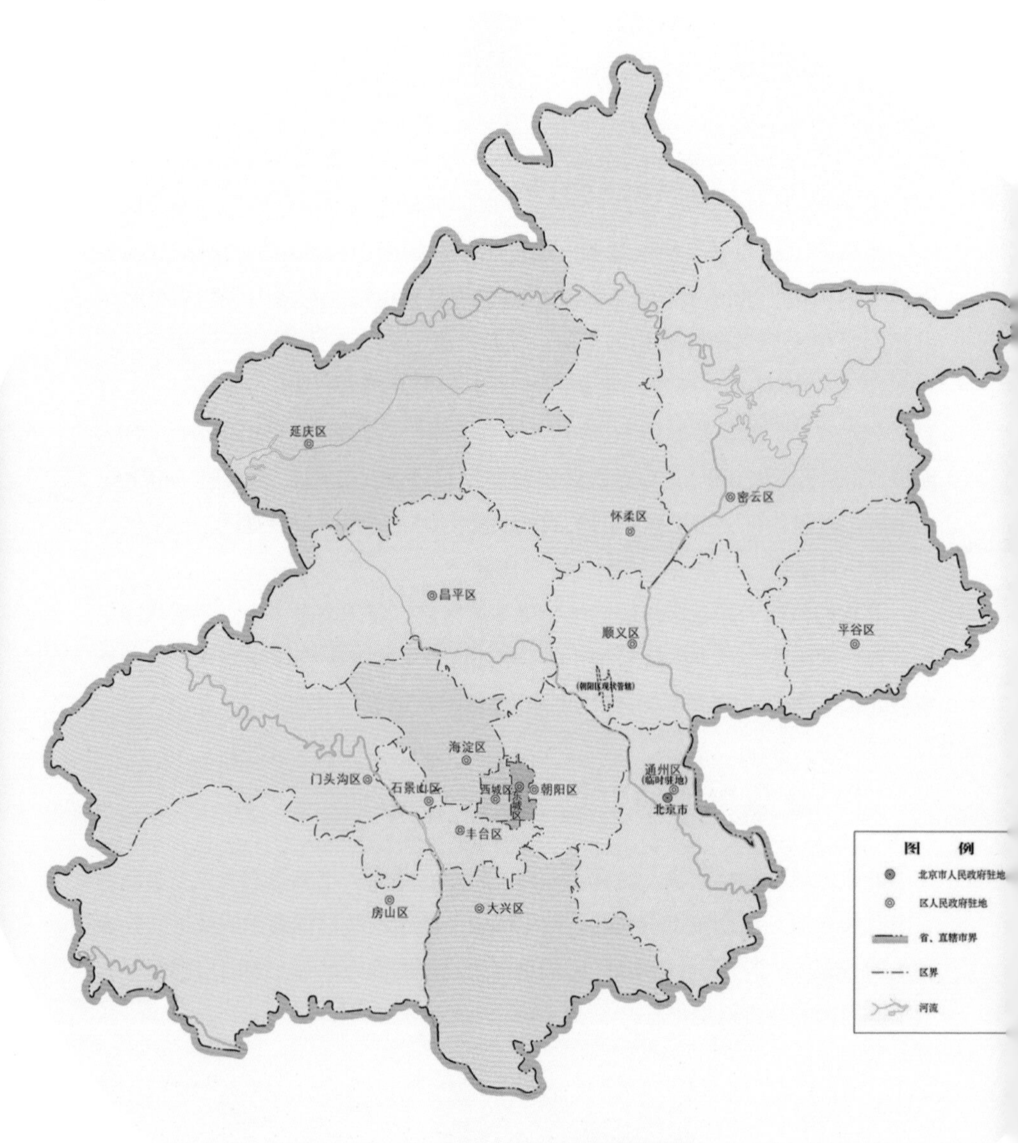
延庆区
怀柔区
密云区
昌平区
顺义区
平谷区
(朝阳区飞地)
海淀区
门头沟区
石景山区
西城区
东城区
朝阳区
通州区
(临时驻地)
北京市
丰台区
房山区
大兴区
图 例
北京市人民政府驻地
区人民政府驻地
省、直辖市界
区界
河流

北京市行政区域图

北京地区的12种陆禽分布广泛，见于各区。褐马鸡 (*Crossoptilon mantchuricum*) 仅见于门头沟区和房山区（小龙门、东灵山、百花山、京西林场）；勺鸡 (*Pucrasia macrolopha*) 见于房山区、门头沟区、昌平区、延庆区、怀柔区、密云区、平谷区海拔600米以上的山地；环颈雉、珠颈斑鸠（*Spilopelia chinensis*）、山斑鸠 (*Streptopelia orientalis*) 见于城市公园、平原和山地；鹌鹑、毛腿沙鸡（*Syrrhaptes paradoxus*）在迁徙季节见于平原草地。

2. 游禽

游禽的脚趾间具蹼（蹼有多种），擅游泳。尾脂腺发达，能分泌大量油脂涂抹于全身羽毛，以保护羽衣不被水浸湿。喙形或扁或尖，适于在水中滤食或啄鱼。代表种类有绿头鸭 (*Anas platyrhynchos*)、䴙䴘、潜鸟、鸬鹚等。

北京地区的79种游禽主要分布于平原和山区的各类湿地。迁徙季节，延庆区（野鸭湖）、密云区（不老屯）、昌平区（沙河水库）会出现种类繁多的游禽，包括天鹅、雁、河鸭、潜鸭、秋沙鸭、麻鸭、鸬鹚、鸥等，䴙䴘除了在上述水域分布外，还于各区有水域的公园繁殖。近年鸳鸯 (*Aix galericulata*) 在怀柔区（怀沙河、怀九河）、门头沟区（百花山）、城市公园（北海、圆明园、颐和园、北京动物园、紫竹院、翠湖）均有分布或繁殖记录。

3. 涉禽

涉禽的外形具有“三长”特征，即喙长、颈长、后肢（腿和脚）长，适于涉水生活。因为后肢长，可以在较深水处捕食和活动。它们趾间的蹼膜往往退化，因此不擅游水。典型的代表种类是鹤和鹭。体形较小但种类繁多的鸻类和鹬类都属于典型的涉禽。

北京地区的96种涉禽主要分布于城郊各区的湿地及有水域的公

园，黄斑苇鳽 (*Ixobrychus sinensis*) 在挺水植物茂密处繁殖，苍鹭 (*Ardea cinerea*)、夜鹭 (*Nycticorax nycticorax*)、池鹭 (*Ardeola bacchus*)、白鹭 (*Egretta garzetta*) 等中大型鹭鸟也在城区周边及远郊地区的湿地和养鱼塘觅食，并在其附近林地的林冠层筑巢繁殖。灰鹤 (*Grus grus*) 喜到延庆区（野鸭湖、官厅水库）、密云区（不老屯）等地区越冬，白枕鹤 (*Grus vipio*) 在春季迁徙时曾集群出现于密云区（陈各庄）、延庆区（野鸭湖）、平谷区（金海湖）。

4. 攀禽

攀禽的脚趾类型发生多种变化，适于在岩壁、石壁、土壁、树干等处行攀缘生活。如两趾向前、两趾朝后的啄木鸟、杜鹃，四趾朝前的雨燕，三、四趾基部并连的戴胜 (*Upupa epops*)、翠鸟等均属于攀禽。

北京地区的 30 种攀禽分布在不同环境区域。常年留居的啄木鸟从城市平原至海拔 1400 米的山地都有分布，它们的栖息地均为林地。夏候鸟普通雨燕 (*Apus apus*) 分布在城郊各区的古建筑及立交桥周围，利用建筑物的孔洞造巢繁殖。夏候鸟大杜鹃 (*Cuculus canorus*) 集中在有东方大苇莺 (*Acrocephalus orientalis*) 繁殖的湿地苇丛地带，四声杜鹃 (*Cuculus micropterus*) 则寻找城市公园及中低山有鸦科鸟类繁殖的地方，它们分别利用苇莺、灰喜鹊 (*Cyanopica cyanus*) 等鸟类完成巢寄生。戴胜一年四季都能见到，它们选择在城市公园、河流湖泊边、低山等地的树洞或柴垛缝隙处造巢繁殖。普通翠鸟 (*Alcedo atthis*) 在城区及城郊各区有水面的地方都能见到，它们捕食水中的小鱼和大型昆虫，在土壁上掘洞营巢繁殖。

5. 猛禽

猛禽的喙、爪锐利带钩，视觉器官发达，飞翔能力强，多捕杀动物为食。羽色较暗淡，常以灰色、褐色、黑色、棕色为主要体色。代表种类有日行性的金雕 (*Aquila chrysaetos*)、雀鹰 (*Accipiter nisus*)、红隼 (*Falco tinnunculus*)、猎隼 (*Falco cherrug*) 和夜行性的雕鸮 (*Bubo bubo*) 等。

北京地区的猛禽，在迁徙季节有鹰形目、隼形目、鸮形目共 51 种飞越北京上空，最集中的迁徙通道当属小西山（太行山余脉至燕山山脉）上空，最多时，日通过量超过千只。各区山地高海拔处有金雕、秃鹫 (*Aegypius monachus*)、雕鸮等常年留居并繁殖，也有苍鹰 (*Accipiter gentilis*)、赤腹鹰 (*Accipiter soloensis*)、日本松雀鹰 (*Accipiter gularis*)、红隼、燕隼 (*Falco subbuteo*)、红脚隼 (*Falco amurensis*)、北领角鸮 (*Otus semitorques*)、红角鸮 (*Otus sunia*) 在海拔 1000 米以下中低山至平原繁殖。城区尚有红隼、纵纹腹小鸮 (*Athene noctua*) 繁殖。冬候鸟大鵟 (*Buteo hemilasius*)、毛脚鵟 (*Buteo lagopus*) 喜在城郊各区的旷野上空盘旋，长耳鸮 (*Asio otus*) 分布于城市公园、平原、低山林地，而短耳鸮 (*Asio flammeus*) 栖息于城郊各区的荒地草丛之中。

6. 鸣禽

鸣禽种类繁多，鸣叫器官（鸣肌和鸣管）发达。它们善于鸣叫，巧于营巢，繁殖时有复杂多变的行为，身体为中、小型，雏鸟均为晚成，在巢中得到亲鸟的哺育才能正常发育。代表种类有喜鹊 (*Pica serica*)、乌鸦、山雀、百灵、家燕、椋鸟、麻雀 (*Passer montanus*) 等。

北京地区鸣禽种类最为丰富，共有 247 种，在全市各区都有许多种类分布。在北京分布的中国特有种共 8 种，除褐马鸡外，其他 7 种均为鸣禽：作为留鸟的山噪鹛（*Pterorhinus davidi*）、银喉长尾山雀（*Aegithalos glaucogularis*）主要分布于各区的中低海拔山地；中华朱雀（*Carpodacus davidianus*）主要分布于北京西部和北部山区海拔 1100 米以上的亚高山林地；乌鸫 (*Turdus mandarinus*) 近 20 年来进入北京，主要分布于城市公园、城区周边的平原地带；黄腹山雀 (*Pardaliparus venustulus*) 繁殖在各区山地阔叶林及针阔混交林，其他季节扩散到平原及城市公园，有些个体到长江以南越冬。夏候鸟宝兴歌鸫 (*Turdus mupinensis*) 在海拔 1200 米以上亚高山林地繁殖。冬候鸟贺兰山红尾鸲 (*Phoenicurus alaschanicus*)，在门头沟（东灵山）

亚高山林缘灌丛地带越冬。此外，喜鹊、灰喜鹊 (*Cyanopica cyanus*)、麻雀等伴人鸟种常年留居于各区的居民区或在民居附近造巢繁殖。

Natural Environment in Beijing

•Nature and Geography of Beijing

Beijing is located at the northwest end of the North China Plain, where the Xishan Mountain of the Taihang Mountains and the Jundu Mountain in the Yanshan Mountains intersect at Nankouguangou, forming a large semicircular mountain bend that opens only southeastward. The area surrounded by this bend, in the southeast part, is a gently sloping fan-shaped plain that extends slowly towards the Bohai Sea, and is also known as "Beijing Bay".

Beijing is adjacent to Tianjin and, together with Tianjin, is surrounded by Hebei Province. It is located at 39°38′-41°05′ N, 115°25′-117°30′ E. It covers a total area is 16,410.54 km^2, with the highest peak being Donglingshan in Mentougou District 2,303 meters high. In 2020, the administrative divisions of Beijing were reorganized into 16 districts: Dongcheng, Xicheng, Chaoyang, Fengtai, Shijingshan, Haidian, Mentougou, Fangshan, Tongzhou, Shunyi, Changping, Daxing, Huairou, Pinggu, Miyun, and Yanqing.

The climate in Beijing belongs to the warm temperate zone, half moist continental monsoon climate, featuring four distinct seasons: hot and rainy summers, cold and dry winters, and short springs and autumns. The average annual temperature ranges between 8 to 12 °C. The average annual precipitation is 600 mm seasonally very uneven, with 70% to 80% of the annual precipitation concentrated between June and August.

•Ecological Groups and Distribution of Birds in Beijing

The birds in Beijing can be sorted into six functional groups: landfowl, waterfowl, wading birds, scansorial birds, raptors and passerines. In mountainous areas, plains, and various habitats in the urban area, each functional group has representative species distributed in various patterns. Due to birds' adeptness at flying, they possess a strong ability to choose habitats, and their distribution often changes with factors such as seasons and food availability.

Administrative division of Beijing

1. Landfowl

Landfowl have robust hind limbs, suitable for walking on the ground; they have short and rounded wings; their bills are strong and often curved, suitable for pecking. Representative species include Common Pheasant (*Phasianu scolchicus*), Japanese Quail (*Coturnix japonica*), etc. Although doves and pigeons are skilled fliers, their primary feeding occurs on the ground, so they are also classified as landfowl.

12 species of landfowl in Beijing are widely distributed across various districts. Brown Eared Pheasants (*Crossoptilon mantchuricum*) are only found in Mentougou and Fangshan Districts (Xiaolongmen, Donglingshan Mountain, Baihuashan Mountain, Beijing West Forest Farm); Koklass Pheasant (*Pucrasia macrolopha*) are observed can be found at elevations above 600 m (Fangshan District, Mentougou District, Changping District, Yanqing District, Huairou District, Miyun District, Pinggu District); Common Pheasants, Spotted Doves (*Spilopelia chinensis*), and Oriental Turtle Doves (*Streptopelia orientalis*), are found in urban parks, plains and mountain areas; Japanese Quail and Pallas's Sandgrouse (*Syrrhaptes paradoxus*) are seen in plain grasslands during migration seasons.

2. Waterfowl

Waterfowl have webbed toes (with various types of webbing) and are skilled swimmers. They have well-developed uropygial glands that secrete a large amount of oil to apply on their feathers, protecting them from getting wet.. Their bills can be flat or pointed, suitable for filtering or catching fish in the water. Representative species groups are geese and ducks (e.g. Mallards: *Anas platyrhynchos*), grebes, loons, and cormorants, etc.

The 79 species of waterfowl in the Beijing area are primarily distributed in various wetlands in plains and mountains. During the migration season, diverse waterfowl species can be found in locations such as Yanqing District (Yeyahu National Wetland Park) , Miyun District (Bulaotun), Changping District (Shahe Reservoir) . These species include swans, geese, ducks, pochards, shelducks, cormorants, gulls, etc. Grebes in addition to being present in the mentioned water bodies, also breed in parks with water features in various districts. In recent years, Mandarin

Ducks (*Aix galericulata*) have been recorded breeding in areas such as Huairou District (Huaisha River, Huaijiu River), Mentougou District (Baihuashan Mountain), and urban parks (to name a few: Beihai Park, Yuanmingyuan Park, Summer Palace, Beijing Zoo, Zizhuyuan Park, Cuihu National Urban Wetland Park).

3. Wading birds

Wading birds are characterized by the "three long" features: long bills, long necks, and long hind limbs (legs and feet), making them well-suited for wading lifestyle. Due to their long hind limbs, they can forage and move in relatively deep water. The interdigital webbing is often reduced, making them less adept at swimming. Typical representative of wading birds include cranes, egrets and herons. The smaller but diverse species of plovers and sandpipers are also typical wading birds.

The 96 species wading birds in the Beijing area are mainly distributed in wetlands of suburban districts and parks with open water. Yellow Bitterns (*Ixobrychus sinensis*) breed in dense upright aquatic plants, while larger herons such as Grey Heron (*Ardea cinerea*), Black-crowned Nightheron (*Nycticorax nycticorax*), Chinese Pond Heron (*Ardeola bacchus*), and Little Egret (*Egretta garzetta*) forage in the wetlands and fish ponds in the city outskirts and in suburban areas. They also nest in the canopy of nearby trees. Common Crane (*Grus grus*) winters in locations like Yanqing District (Yeyahu National Wetland Park, Guangting Reservoir) and Miyun District (Bulaotun). During spring migration, the White-naped Crane (*Grus vipio*) has been recorded in Miyun District (Chengezhuang), Yanqing District (Yeyahu National Wetland Park) and Pinggu District (Jinhai Lake).

4. Scansorial birds

Scansorial birds have various adaptations in feet structure, suitable for climbing lifestyle on surfaces such as cliffs, rock walls, soil walls, tree trunks, etc. Examples of scansorial birds include woodpeckers and cuckoos with two toes forward and two toes backwards; swifts with four toes forward; and hoopoes and kingfishers with three or four toes fused at the base.

The 30 species of scansorial birds in Beijing are distributed across different habitats. Resident woodpeckers can be found in habitats ranging from urban plains to mountains areas up to an altitude of 1,400m. Their habitats are typically forests.The summer migratory bird, the Common Swift (*Apus apus*), is distributed around ancient buildings and overpasses in various suburb districts of the city, utilizing the cavities in structures for nesting and breeding. The Common Cuckoo (*Cuculus canorus*) occurs in wetland reed areas where Oriental Reed Warbler (*Acrocephalus orientalis*) breeds. The Indian Cuckoo (*Cuculus micropterus*) locates in urban parks and low mountains where corvid birds breed. They brood-paratize birds like the warblers and Azure-winged Magpie (*Cyanopica cyanus*). Eurasian Hoopoe (*Upupa epops*) can be observed throughout the year. They choose tree holes or crevices in woodpiles near urban parks, rivers lakes and low mountains for nesting. The Common Kingfisher (*Alcedo atthis*) is found in areas with water surfaces both urban or suburban, where they prey on small fish and large insects in the water, excavating holes in earthen walls for nesting.

5. Raptors

Raptors have sharp hooked bills and claws, well-developed visual organs, strong flying abilities, and primarily hunt and kill animals for food. They often have dull feather colors, with grey, brown, and black bedding predominant. Representative species include diurnal raptors such as Golden Eagle (*Aquila chrysaetos*), Eurasian Sparrow Hawk (*Accipiter nisus*), Common Kestrel (*Falco tinnunculus*), Saker Falcon (*Falco cherrug*), and nocturnal raptors like Eurasian Eagle Owl (*Bubo bubo*).

In Beijing during the migration season, a total of 51 species of raptors belonging to the orders Accipitriformes, Falconiformes, and Strigiformes fly over the city. The most concentrated migration route is over the Xishan Mountain of the Taihang Mountains to the Yanshan Mountain Range, where, at peak times, the daily migration count exceeds a thousand individuals. In mountainous areas of various districts, at high elevations, species like the Golden Eagle, Cinereous Vulture (*Aegypius monachus*), Eurasian Eagle Owl reside year-round and breed. Below 1,000 m in mid-low mountains to plains, species like Northern Goshawk (*Accipiter gentilis*), Chinese Sparrowhawk (*Accipiter soloensis*), Japanese Sparrowhawk (*Accipiter gularis*), Common Kestrel, Hobby (*Falco subbuteo*), Eastern Red-footed Falcon (*Falco amurensis*), Japanese Scops Owl (*Otus semitorques*), and Oriental Scops Owl (*Otus sunia*) breed. In urban areas, species like Common Kestrel and Little Owl (*Athene noctua*) also breed. Wintering birds, such as Upland Buzzard (*Buteo hemilasius*) and Roughlegged Buzzard (*Buteo lagopus*) prefer to hover over open fields in various suburban districts. The Long-eared Owl (*Asio otus*) is found in city parks, plains, and low mountain forests, while the Short-eared Owl (*Asio flammeus*) inhabits the wastelands and grassy areas of suburban districts.

6. Passerines

Passerines are diverse in species, with well-developed vocal organs (syrinx and song muscles). They are adept in singing, skilled in nest building, and exhibit complex behaviors during reproduction. They are generally medium to small-sized birds, and their chicks are altricial, relying on parental care in the nest for nurturing. Representative species include magpies, crows, tits, larks, swallows, starlings and sparrows, etc.

Beijing is home to a rich variety of passerines, totaling 247 species, distributed across various districts in the city. In Beijing among the 8 bird species endemic to China, with the Brown Eared Pheasant being the exception, the remaining 7 are all passerines. Resident species like the Plain Laughingthrush (*Pterorhinus davidi*) and Silver-throated Bushtit (*Aegithalos glaucoguls*) are mainly found in mid-low elevation mountainous areas across different districts. Chinese Beautiful Rosefinch (*Carpodacus davidianus*) is primarily distributed in the western and northern mountainous areas of Beijing at elevations above 1,100 m; Blackbird (*Turdus mandarinus*) which dispersed into Beijing in the past 20 years, is mainly distributed in urban parks and plains surrounding the city.

In addition, Yellow-bellied Tit (*Pardaliparus venustulus*) breeds in broad-leaved forests and mixed coniferous-broadleaf forests in mountainous areas of all districts. It can disperse to plains and urban parks, with some individuals wintering south of the Yangtze River. The summer breeder Chinese Thrush (*Turdus mupinensis*), breeds in subalpine forests at elevations above 1,200 m. The Winter visitor Alashan Redstart (*Phoenicurus alaschani*cus), winters in shrublands at the edge of subalpine forests in Mentougou District (Dongling Mountain). In addition, synanthropic birds like Oriental Magpie (*Pica serica*), Azure-winged Magpie, and Eurasian Tree Sparrow (*Passer montanus*) reside year-round in residential areas across various districts or nest and breed near human settlements.

北京鸟类区系特点

北京市园林绿化局公布的2023年版北京市陆生野生动物种类已经由596种增加至608种，鸟类达到515种。鸟类组成非常复杂，虽然以古北界种类为主，但也有不少东洋界种类和广布种类，表现出古北界与东洋界物种的交错与混合，且鸟类的分布型也体现出两界及其亚界之间的渗透与过渡。根据张荣祖（2011）《中国动物地理》，按照动物地理区划分级，北京属古北界东北亚界华北区（内含黄土高原亚区和黄淮平原亚区）。

北京地区分布的鸟种主要是北方种类：

古北型的大鸨（*Otis tarda*）、灰鹤（*Grus grus*）、普通雨燕（*Apus apus*）、大杜鹃（*Cuculus canorus*）、普通翠鸟（*Alcedo atthis*）、普通鵟（*Buteo japonicus*）、黑头䴓（*Sitta villosa*）、麻雀（*Passer montanus*）、灰喜鹊（*Cyanopica cyanus*）、喜鹊（*Pica serica*）、秃鼻乌鸦（*Corvus frugilegus*）等；

东北型的白枕鹤（*Grus vipio*）、白头鹤（*Grus monacha*）、白眉姬鹟（*Ficedula zanthopygia*）、金翅雀（*Chloris sinica*）、三道眉草鹀（*Emberiza cioides*）等；

华北型的褐马鸡（*Crossoptilon mantchuricum*）、山噪鹛（*Pterorhinus davidi*）等；

东北型中的广布种鸭、雁和鹭类以及灰斑鸠（*Streptopelia decaocto*）、珠颈斑鸠（*Spilopelia chinensis*）、戴胜（*Upupa epops*）、北领角鸮（*Otus semitorques*）、银喉长尾山雀（*Aegithalos glaucogularis*）等；

东北—华北型的牛头伯劳（*Lanius bucephalus*）、红尾伯劳（*Lanius cristatus*）、灰椋鸟（*Spodiopsar cineraceus*）等。

另外还有少量中亚亚界的蒙新区和青藏区的种类出现，如中亚型的石鸡（*Alectoris chukar*）、斑翅山鹑（*Perdix dauurica*）、毛腿沙鸡

（*Syrrhaptes paradoxus*）、大鵟（*Buteo hemilasius*）、凤头百灵（*Galerida cristata*）、山鹛（*Rhopophilus pekinensis*）等；

喜马拉雅—横断山区型的棕腹啄木鸟（*Dendrocopos hyperythrus*）、中华朱雀（*Carpodacus davidianus*）、蓝鹀（*Emberiza siemsseni*）等；

高地型的斑头雁（*Anser indicus*）、棕头鸥（*Chroicocephalus brunnicephalus*）、红腹红尾鸲（*Phoenicurus erythrogastrus*）、褐岩鹨（*Prunella fulvescens*）、粉红胸鹨（*Anthus roseatus*）等。

北京地区分布的鸟种也有一些南方种类：

东洋型的牛背鹭（*Bubulcus ibis*）、红翅凤头鹃（*Clamator coromandus*）、噪鹃（*Eudynamys scolopaceus*）、大鹰鹃（*Hierococcyx sparverioides*）、四声杜鹃（*Cuculus micropterus*）、黑翅鸢（*Elanus caeruleus*）、赤腹鹰（*Accipiter soloensis*）、红角鸮（*Otus sunia*）、黑枕黄鹂（*Oriolus chinensis*）、发冠卷尾（*Dicrurus hottentottus*）等；

南中国型的勺鸡（*Pucrasia macrolopha*）、黄腹山雀（*Pardaliparus venustulus*）、白头鹎（*Pycnonotus sinensis*）等。

由于北京地区在候鸟迁徙的通道上，每年春、秋两季，都会有种类繁多的雁鸭、鸥、鸻鹬、猛禽以及燕科、柳莺科、鸫科、鹟科、鹡鸰科、燕雀科、鹀科等雀形目鸟类出现在北京，使得北京鸟类的种类非常丰富。

Characteristics of Beijing Bird Fauna

According to the 2023 announcement of the list of terrestrial vertebrate species by Beijing Municipal Forestry and Parks Bureau, the total number of species has increased from 596 to 608, with 515 species being birds. The bird species composition in Beijing is highly complex, although the predominant species belong to the Palearctic realm, there is also a significant presence of species from the Oriental realm and widespread species, indicating the interaction and transition between the two realms and their sub-realms. According to *China Zoological Geography* (Zhang, 2011), following the classification of animal geographic zones, Beijing falls within the Palaearctic realm, Northeast sub-realm, and North China region, which includes the Loess Plateau sub-region and the Yellow-Huai Plain sub-region.

The bird species Beijing fauna mainly consist of are Northern types:

Palearctic type: Great Bustard (*Otis tarda*), Common Crane (*Grus grus*), Common Swift (*Apus apus*), Common Cuckoo (*Cuculus canorus*), Common Kingfisher (*Alcedo atthis*), Eastern Buzzard (*Buteo japonicus*), Chinese Nuthatch (*Sitta villosa*), Eurasian Tree Sparrow (*Passer montanus*), Azure-winged Magpie (*Cyanopica cyanus*), Oriental Magpie (*Pica serica*), Rook (*Corvus frugilegus*), etc.

Northeast type: White-naped Crane (*Grus vipio*), Hooded Crane (*Grus monachal*), Yellow-rumped Flycatcher (*Ficedula zanthopygia*), Oriental Greenfinch (*Chloris sinica*), Meadow Bunting (*Emberiza cioides*), etc.

North type: Brown Eared Pheasant (*Crossoptilon mantchuricum*), Plain Laughingthrush (*Pterorhinus davidi*), etc.

Widespread species in Northeast type: Ducks, geese, egrets and herons, Eurasian Collared Dove (*Streptopelia decaocto*), Spotted Dove (*Spilopelia chinensis*), Eurasian Hoopoe (*Upupa epops*), Japanese Scops Owl (*Otus semitorques*), Silver-throated Bushtit (*Aegithalos glaucogularis*), etc.

Northeast-North type: Bull-headed Shrike (*Lanius bucephalus*), Brown Shrike (*Lanius cristatus*), White-cheeked Starling (*Spodiopsar cineraceus*), etc.

In addition, there are a few species from the Central Asian realm and the Qinghai-Xizang Plateau also appear in Beijing. e.g. Central Asia type: Chukar Partridge (*Alectoris chukar*), Daurian Partridge (*Perdix dauurica*), Pallas's Sandgrouse (*Syrrhaptes paradoxus*), Upland Buzzard (*Buteo hemilasius*), Crested Lark (*Galerida cristata*), Chinese Hill Babbler (*Rhopophilus pekinensis*), etc. Himalayas-Hengduan Mountains type: Rufous-bellied Woodpecker (*Dendrocopos hyperythrus*), Chinese Beautiful Rosefinch (*Carpodacus davidianus*), Slaty Bunting (*Emberiza siemsseni*), etc. Highland type: Bar-headed Goose (*Anser indicus*), Brownheaded Gull (*Chroicocephalus brunnicephalus*), White-winged Redstart (*Phoenicurus erythrogastrus*), Brown Accentor (*Prunella fulvescens*), Rosy Pipit (*Anthus roseatus*), etc.

Beijing also hosts some Southern types:

Oriental type: Cattle Egret (*Bubulcus ibis*), Chestnut-winged Cuckoo (*Clamator coromandus*), Common Koel (*Eudynamys scolopaceus*), Large Hawk-Cuckoo (*Hierococcyx sparverioides*), Indian Cuckoo (*Cuculus micropterus*), Black-winged Kite (*Elanus caeruleus*), Chinese Sparrowhawk (Accipiter soloensis), Oriental Scops Owl (*Otus sunia*), Black-naped Oriole (*Oriolus chinensis*), Hair-crested Drongo (*Dicrurus hottentottus*), etc.

Southern China type: Koklass Pheasant (*Pucrasia macrolopha*), Yellowbellied Tit (*Pardaliparus venustulus*), Light-vented Bulbul (*Pycnonotus sinensis*), etc.

Due to Beijing's location along the East Asian-Australasian flyway of birds, a diverse array of waterfowl, gulls, terns, raptors, as well as species from the Hirundinidae, Phylloscopidae, Turdidae, Muscicapidae, Motacillidae, and Fringillidae appear in Beijing during the spring and autumn seasons, contributing to the rich diversity of bird species in this region.

常见鸟种
Common Birds

中文名 **环颈雉（雉鸡）**
学　名 *Phasianus colchicus*
英文名 Common Pheasant

鸡形目
GALLIFORMES

雉科
Phasianidae

环颈雉是一种大型陆禽。雌雄差异大。雄鸟羽色华丽，具金属光泽，头部主要为金属蓝绿色光泽，头顶偏白，脸部具红色裸皮，多数亚种具明显的白色“围脖”，尾羽较长，可达50余厘米。雌鸟外貌较为朴素，全身多为黄褐色，具深棕色杂斑。

环颈雉常栖息于草木茂盛的山地、沼泽地、农田等环境。如遇到危险，会趴卧在原地，依靠保护色很好地隐藏于杂草、灌木丛中。春季求偶季节，雄性环颈雉会站在农田土埂上，扯着脖子发出单调而响亮的“嘎，嘎”声。主要以植物种子和果实为食。

环颈雉在我国的多数地区都有分布。北京地区的延庆、密云、怀柔等地较为常见，为留鸟。

北京市重点保护 Beijing priority conservation	国家重点保护 State priority conservation	IUCN
——	——	无危（LC）2016.10.1

雌鸟 Female

Common Pheasant. is a large landfowl with distinct sexual dimorphism. Male has gorgeous, iridescent plumage; head primarily metallic blue and green; whitish crown; face with red bare part; most subspecies have distinctive white collar; elongated tail reaches more than 50cm. Female duller, mostly tawny with dark brown spots.

Often inhabits environments such as mountainous areas with dense vegetation, marshes, and agricultural fields. When in danger, it will lies on the ground and remains camouflaged in grasses and shrubs. During the breeding season, male produces loud and monotonous "quack, quack" while standing on dirt mounds with a stretched neck. Feeds mostly on seeds and fruits.

Occurs in most regions in China. A resident in Beijing where it is common in suburban areas such as Yanqing, Miyun, and Huairou.

中文名 **岩鸽**
学　名 *Columba rupestris*
英文名 Hill Pigeon

鸽形目
COLUMBIFORMES

鸠鸽科
Columbidae

岩鸽是一种中型陆禽。羽色及体型甚似家鸽。雌雄相似。头部灰色，颈部具金属绿色光泽，胸部沾粉色，腹部白色，背部灰色，两翼具两道黑色的不完整翼斑，尾羽灰色，具一道明显的白色次端斑。

岩鸽常栖于多石的山区地带。常成群活动，有时可见百只的大群。主要以草籽和植物果实为食，有时也吃昆虫等。

岩鸽在我国主要分布于北方及西南地区。北京较常见于房山、怀柔等山区，为留鸟。

北京市重点保护 Beijing priority conservation	国家重点保护 State priority conservation	IUCN
是（YES）	——	无危（LC）2016.10.1

成鸟 Adult

A medium-sized pigeon. Plumage and shape similar to domestic pigeon. Sexes similar. Gray head and back, neck metallic green, pinkish chest and white belly; two incomplete black wing bars on each wing; gray tail with one distinct white subterminal band.

Often inhabits rocky, mountainous areas. Often travel in flocks, sometimes up to 100 individuals. Feeds mostly on seeds and fruits, but sometimes on insects as well.

Mostly occurs in northern and southwestern China. A resident in Beijing where it is common in mountainous areas such as Fangshan and Huairou.

中文名 山斑鸠

学　名 *Streptopelia orientalis*

英文名 Oriental Turtle Dove

鸽形目 COLUMBIFORMES

鸠鸽科 Columbidae

山斑鸠是北方常见的斑鸠中羽色比较漂亮的一种。雌雄相似。头部偏灰色，胸腹部灰粉色，背部和两翼棕褐色。最明显的识别点是颈侧具黑色及白色相间的横纹。

山斑鸠常栖息于山区林地、郊区村镇，有时在城区公园中可以见到。主要以植物的果实、嫩叶为食，有时也吃昆虫等小型动物。飞行能力强，繁殖期常张翼在空中盘旋炫耀飞行。

山斑鸠广泛分布于我国各地。在北京各地几乎都有分布，为留鸟。

北京市重点保护 Beijing priority conservation	国家重点保护 State priority conservation	IUCN
——	——	无危（LC）2016.10.1

A rather beautiful pigeon compared to other common ones in northern China, sexes similar. Grayish head, pinkish gray underparts; brown back and wings; diagnostic black-and-white striped patch on neck.

Often inhabits woodland, suburb, sometimes urban parks. Feeds mostly on fruits and young leaves, but sometimes also on small animals such as insects. Strong flight ability, often seen soaring flamboyantly in the sky during the breeding season.

Widely distributed in all regions of China. A resident in Beijing where it occurs almost everywhere.

中文名 **灰斑鸠**
学　名 *Streptopelia decaocto*
英文名 Eurasian Collared Dove

鸽形目
COLUMBIFORMES
鸠鸽科
Columbidae

成鸟 Adult

灰斑鸠是一种小型陆禽。雌雄相似。体型较家鸽略小。全身羽毛以灰色为主，胸腹部颜色浅，背部及两翼略深。颈部具显著的黑色横纹。

灰斑鸠生活在开阔平原地带的农田、耕地等环境。主要以草籽和植物果实为食，有时也吃昆虫等小型动物。

灰斑鸠在我国主要分布于华北、西北以及长江中下游地区。北京较常见于密云、延庆等地，为留鸟。

北京市重点保护 Beijing priority conservation	国家重点保护 State priority conservation	IUCN
——	——	无危（LC）2019.8.14

A small pigeon, sexes similar, slightly smaller than a domestic pigeon. Body feathers mostly gray; chest and belly light-colored; back and wings slightly darker; distinct black collar on neck.

Inhabits agricultural fields in plains. Feeds mostly on seeds and fruits, but sometimes also on small animals such as insects.

Occurs mostly in north and northwestern China, and the middle and lower reaches of the Yangtze River. A resident in Beijing where it is common in places such as Miyun and Yanqing.

中文名 **珠颈斑鸠**
学　名 *Spilopelia chinensis*
英文名 Spotted Dove

鸽形目
COLUMBIFORMES

鸠鸽科
Columbidae

珠颈斑鸠是我国分布最广、最为常见的一种斑鸠。雌雄相似。头部偏灰色，颈部沾粉色，胸腹部及两翼多为灰褐色。它得名“珠颈”的原因在于颈部具黑底、白色点状斑纹，似珍珠。尾较山斑鸠长。

珠颈斑鸠适宜各类生境，山区林地、城市公园、居民小区、学校都能见到它们的身影。主要以草籽等植物为食，偶尔也捕捉昆虫。不甚惧人，有时甚至在阳台的花盆中繁殖。春季，常在小区中听到其“咕咕—咕”的叫声。

珠颈斑鸠在中国东部地区分布较广，在北京各地都有分布，为留鸟。

北京市重点保护 Beijing priority conservation	国家重点保护 State priority conservation	IUCN
——	——	无危（LC）2016.10.1

Most common and widely distributed wild pigeon species in China, sexes similar. Grayish head, pinkish neck, underparts and wings grayish brown; named “pearl neck” in Chinese because of the black-and-white neck patch, which resembles pearls; tail longer than that of Oriental Turtle Dove.

Well-adapted to all kinds of habitats including mountainous woodlands, urban parks, residential areas, and schools. They mostly feed on seeds but sometimes insects as well. Unafraid of people, sometimes even build nests in flowerpots on the balcony. In spring, its “gugu, gu” is often heard in residential areas.

Widely distributed in eastern China. A common resident bird all across Beijing.

006 中文名 大天鹅

学　名 *Cygnus cygnus*

英文名 Whooper Swan

雁形目
ANSERIFORMES

鸭科
Anatidae

大天鹅是一种大型游禽。外形与小天鹅很相似。雌雄相似。其体型略大于小天鹅，颈部更长更细，与小天鹅最主要的区别在喙部，大天鹅的喙部黄色面积大且过鼻孔，黄色、黑色相交处为斜面。亚成鸟体色为灰褐色，喙沾粉色。

大天鹅是鸟类中少见的实行“一夫一妻制”的鸟类。也是世界上罕见的可以飞跃珠穆朗玛峰的鸟类，迁徙时它们的飞行高度最高可达9000米。习性似小天鹅，通常较为惧人。常栖息于湖泊、水库、河流中，主要以水生植物为食。

大天鹅在我国的西北部、东部、南部等地都有分布。北京可见于郊区，迁徙时一些城市湿地公园里可见。通常春季迁徙时间较小天鹅更早，秋季较晚，为旅鸟和冬候鸟。

北京市重点保护 Beijing priority conservation	国家重点保护 State priority conservation	IUCN
—	二级（II）	无危（LC）2016.10.1

亚成鸟 Immature

A large waterfowl, sexes similar. Shape similar to Tundra Swan, but slightly larger and has more slender neck; best distinguished by large yellow patch that goes past the nostrils and separated diagonally from the black bill. Subadult grayish brown with pinkish bill.

Monogamous mating system. Among the few birds that can fly over Mount Qomolangma; can fly as high as 9000m during migration. Behaves like Tundra Swan; normally afraid of people. Often inhabits lakes, reservoirs, and rivers. Feeds mostly on aquatic plants.

Occurs in regions such as northwestern, eastern, and southern China. A passage migrant and winter visitor in Beijing where it is found in suburban areas, and urban wetland parks during migration. Compared to Tundra Swan, Whooper Swan migrates earlier in spring but later in fall.

中文名 **小天鹅**
学　名 *Cygnus columbianus*
英文名 Tundra Swan

雁形目
ANSERIFORMES

鸭科
Anatidae

成鸟 Adult

小天鹅是一种大型游禽。通体白色，雌雄相似。成鸟整体白色，喙部颜色前黑后黄（黄斑面积小，不过鼻孔）。跗跖黑色。亚成鸟体色为灰褐色，喙沾粉色。

小天鹅在迁徙及越冬时期常成群活动。在水中游动时颈部较为挺拔。通常较为惧人，警戒距离远。主要以水生植物为食，有时亦吃农作物种子或捕食螺类、鱼虾等。

小天鹅可见于我国东部多数地区。京郊的怀柔水库、密云水库，甚至颐和园等地可见迁徙种群，为旅鸟和少见冬候鸟。

北京市重点保护 Beijing priority conservation	国家重点保护 State priority conservation	IUCN
——	二级（II）	无危（LC）2016.10.1

亚成鸟 Immature

A large waterfowl, sexes similar. Adult overall white; black bill with small yellow patch that does not go past the nostrils; black tarsi. Subadult grayish brown; bill tinged pinkish.

Form flocks during migration and winter. Neck held relatively upright when swimming. Normally afraid of people; long flight initiation distance. Feeds mostly on aquatic plants, but also seeds from crops, snails, fish, and shrimps.

Occurs in most regions in eastern China. A passage migrant and winter visitor found at Huairou and Miyun Reservoirs, and even at the Summer Palace.

008

中文名 **短嘴豆雁**
学　名 *Anser serrirostris*
英文名 Tundra Bean Goose

雁形目 ANSERIFORMES
鸭科 Anatidae

短嘴豆雁是一种体型较大的游禽。雌雄相似。整体多灰褐色、棕褐色羽毛，最显著的特点是黑色喙部近尖端具一黄色斑点，形如黄豆。胸腹部偏灰白色，尾羽灰褐色，跗跖橙红色。

短嘴豆雁常集群活动，多时可达千只。飞行时集成“人”字形或“一”字形的雁阵长途行进。主要以水中、岸边的水草以及农田中的谷物等为食。

短嘴豆雁在我国的多数地区都有分布。北京可见于郊区，偶见于城区公园，为旅鸟和冬候鸟。

北京市重点保护 Beijing priority conservation	国家重点保护 State priority conservation	IUCN
—	—	无危（LC）2018.8.9

成鸟 Adult

A relatively large waterfowl, sexes similar. Plumage mostly brownish; distinct orange patch near the tip of the black bill; breast and belly gray to white; tail feathers grayish brown; orange tarsi.

Often form flocks, sometimes up to one thousand individuals. Forming either a “V” or a horizontal line when moving long distances. Feeds mostly on aquatic plants in the water and on the riverbank, and crops in farmland.

Occurs in most regions of China. A passage migrant and winter visitor in Beijing where it is common in suburbs and urban parks.

009

中文名 **鹊鸭**
学　名 *Bucephala clangula*
英文名 Common Goldeneye

雁形目
ANSERIFORMES
鸭科
Anatidae

鹊鸭是一种体型中等的游禽。形状较为特殊，头顶高高鼓起。雌雄差异大。雄鸟头部带金属光泽的绿色羽毛，在阳光不足时发黑色，喙基有一圆形白斑，虹膜亮黄色。雌鸟羽毛较暗淡，头部棕色。

鹊鸭常栖于水库、湖泊、河流中，喜在流速较慢的水域活动。常成群活动，时而下潜觅食鱼虾、昆虫等。

鹊鸭广泛分布于我国各地。北京地区的郊区水库地带遇见率较高，偶在城区公园如颐和园可见，为冬候鸟和旅鸟。

北京市重点保护 Beijing priority conservation	国家重点保护 State priority conservation	IUCN
是（YES）	——	无危（LC）2018.8.7

雌鸟 Female

A medium-sized waterfowl. Special shape with large head in an upright posture, sexes different. Head metallic green in male, but black in poor lighting; a circular white patch near the bill; bright yellow iris; female duller with brown head.

Often inhabits reservoirs, lakes, and rivers; prefers slow-moving water. Form flocks. Sometimes dives to feed on fish, shrimps, and insects.

Widely distributed in China. A winter visitor and passage migrant in Beijing where it is mostly found at suburban reservoirs but sometimes in urban parks including the Summer Palace.

中文名 斑头秋沙鸭（白秋沙鸭）

学　名 *Mergellus albellus*

英文名 Smew

雁形目 ANSERIFORMES

鸭科 Anatidae

斑头秋沙鸭为小型游禽。体长约40厘米。雌雄差异大。雄鸟繁殖羽体羽由黑白两色构成，喙为灰色，两胁沾灰色细纹。雌鸟整体羽色暗淡，头顶棕红色，喉部白色，体羽多灰色。

斑头秋沙鸭常栖息于湖泊、水库、河流等环境。常成小群活动，善游泳及潜水。食性较杂，主要以水草、鱼虾为食。较惧人，一般难以接近。

斑头秋沙鸭在我国多数地区都有分布，但总数不多。北京郊区的水库、河流中可见，在颐和园有稳定越冬记录，为旅鸟和冬候鸟。

北京市重点保护 Beijing priority conservation	国家重点保护 State priority conservation	IUCN
—	二级（II）	无危（LC）2016.10.1

雌鸟 Female

A small waterfowl, body length about 40cm, sexes different. Breeding male has black and white plumage, gray bill, grayish barring on flanks. Female duller, with rufous-brown crown and white throat; body feathers mostly gray.

Often inhabits environments such as lakes, reservoirs, and rivers. Form small flocks. Adept at swimming and diving. Feeds mostly on aquatic plants, fish, and shrimps. Rather afraid of people, difficult to approach.

Occurs in most regions in China, but total population relatively low. A passage migrant and winter visitor in Beijing where it is found in suburban reservoirs and rivers, also reliably found at the Summer Palace during winter.

中文名 **普通秋沙鸭**
学　名 *Mergus merganser*
英文名 Common Merganser

雁形目
ANSERIFORMES
鸭科
Anatidae

普通秋沙鸭体型较大，体长近 70 厘米。雌雄差异大。雄鸟繁殖羽头部具绿色金属光泽，喙红色、尖端带钩，胸腹部白色，背部主要为绿色。雌鸟头为棕褐色，喙红色，头上常具不明显羽冠，胸腹较白，背和两翼多灰色。

普通秋沙鸭常栖于河流、湖泊、水库等地带。常成群活动。主要以鱼虾等小型动物为食。

普通秋沙鸭广泛分布于我国各地。北京郊区的水库、湖泊中不难见到，有时在城区公园中也可见，为旅鸟和冬候鸟。

保护级别 Protected-level	北京市重点保护 Beijing priority conservation	国家重点保护 State priority conservation	IUCN
	是（YES）	——	无危（LC）2018.8.7

A rather large waterfowl, body length almost 70cm, sexes different. Breeding male has green metallic head; red, hooked bill; white underparts; back mostly green. Female has rusty cinnamon head, red bill, inconspicuous crest, and white underparts; back and wings mostly gray.

Often inhabits rivers, lakes, and reservoirs. Form flocks. Feeds mostly on fish, shrimps, and other small aquatic animals.

Widely distributed in China. A passage migrant and winter visitor in Beijing where it is found in suburban reservoirs and lakes, as well as urban parks.

012 中文名 赤麻鸭

学　名 *Tadorna ferruginea*

英文名 Ruddy Shelduck

雁形目 ANSERIFORMES

鸭科 Anatidae

赤麻鸭是雁鸭类中非常容易辨认的大型游禽，身长近 70 厘米。雌雄相似。全身羽毛以棕黄色为主，头部、背部略浅，颜色偏白，喙黑色。尾尖、翼尖等部位有黑色的羽毛。雄鸟颈部常具一黑色颈环。

赤麻鸭适应能力强，低海拔到高海拔的草原、湖泊、河流、农田甚至城市公园的湖泊等环境都能生活。水生植物、昆虫、贝类、鱼虾都是其食物。

赤麻鸭广泛分布于我国各地，也是北京地区最常见的鸭科鸟类之一，在各类水域环境都有机会看到，为旅鸟和冬候鸟。

北京市重点保护 Beijing priority conservation	国家重点保护 State priority conservation	IUCN
是（YES）	——	无危（LC）2016.10.1

雌鸟 Female

One of the most distinctive large waterfowl in Beijing, body length nearly 70cm, sexes similar. Plumage mostly ruddy; head and back paler; black bill; tail and wings have dark tips. Male often has dark neck ring.

Adapt to a range of environments from low to high elevations: steppes, lakes, rivers, farmland, and even urban parks. Feeds on aquatic plants, insects, shellfish, and fish.

Widely distributed in China. A passage migrant and winter visitor in Beijing where it is one of the most common anatids; found in all kinds of aquatic environment.

中文名 **鸳鸯**
学　名 *Aix galericulata*
英文名 Mandarin Duck

雁形目
ANSERIFORMES
鸭科
Anatidae

鸳鸯是一种广为人知的游禽。雌雄差异大。雄鸟繁殖羽鲜艳多彩，喙为红色，额头绿色，头部具一宽阔的白色眉纹，颈部棕红色，两翼具一特化的棕色帆状羽。雌鸟全身灰褐色，喙深褐色，具白色眼圈，眼后具一白色眉纹，两胁具偏白色点斑。

在我国传统文化中，鸳鸯是最受人们喜爱的鸟类之一，常被当作美好爱情的象征。但现代研究发现，在繁殖期内，雄鸟和配偶交配后，通常会离开去寻找其他雌鸟，并不承担抚育后代的责任。它们栖息在淡水湖泊、池塘中，以水中植物、昆虫、鱼虾等为食。

鸳鸯在我国东部、中部地区广泛分布。北京郊区和几处城市公园中可见，为夏候鸟和留鸟。

北京市重点保护 Beijing priority conservation	国家重点保护 State priority conservation	IUCN
——	二级（II）	无危（LC）2018.8.9

雌鸟 Female

A well-known waterfowl, Sexes different. Breeding male colorful; red bill; green forehead; one broad supercilium; reddish brown neck, wings have specialized brown sail-like feathers. Female mostly gray; bill grayish brown; white eye-ring; white eye-stripe behind the eye; flanks with white spots.

A beloved bird in Chinese culture, symbolizing love. However, recent research shows that male looks for other females after mating with one, without fulfilling any parental roles. Inhabits freshwater lakes and ponds. Feeds on aquatic plants, insects, fish, and shrimps.

Widely distributed in eastern and central China. A migratory breeder and resident in Beijing where it is found in suburbs and several urban parks.

014 中文名 风头潜鸭

学　名 *Aythya fuligula*

英文名 Tufted Duck

雁形目 ANSERIFORMES

鸭科 Anatidae

风头潜鸭为中型游禽，体长近50厘米。雌雄差异大。雄鸟繁殖羽头部紫黑色，具明显羽冠，喙偏灰色，喙尖黑色，胸部偏黑色，腹部白色，背部、两翼黑褐色。雌鸟整体羽色暗淡，多棕褐色。

风头潜鸭常栖息于开阔的湖泊、水库等环境。常成群活动，善游泳及潜水。食性杂，水草、鱼虾、贝类都是其食物。

风头潜鸭在我国多数地区都有分布。北京郊区水库、湖泊中常见，偶在颐和园、圆明园等城区公园中出现，为旅鸟。

北京市重点保护 Beijing priority conservation	国家重点保护 State priority conservation	IUCN
是（YES）	——	无危（LC）2016.10.1

雌鸟 Female

A medium-size waterfowl, body length nearly 50cm, sexes different. Breeding male has dark purple head with distinct crest; black tip on grayish bill; black chest; lower belly white; blackish back and wings. Female duller and mostly brown.

Often inhabits environments such as open lakes and reservoirs. Often form flocks. Adept at swimming and diving. Omnivorous diet consists of aquatic plants, fish, shrimps, and shellfish.

Occurs in most regions in China. A passage migrant in Beijing where it is often found in suburban reservoirs and lakes. Sometimes seen at urban parks such as the Summer Palace and Yuanmingyuan.

中文名 赤膀鸭

学　名 *Mareca strepera*

英文名 Gadwall

雁形目
ANSERIFORMES

鸭科
Anatidae

赤膀鸭是一种体色比较素雅的游禽。雌雄相似。雄鸟繁殖羽整体灰色，胸部具黑色斑纹，喙黑色。雌鸟整体黄褐色，与绿头鸭雌鸟很相似。赤膀鸭外表素雅，与其他鸭科鸟类很难区分，相对容易识别的特点是翼镜为亮白色。

赤膀鸭喜欢栖息在湖泊、水库、沼泽等内陆水域中，尤其喜欢在富有水生植物的开阔水域活动。常成小群活动。主要以水生植物为食。

赤膀鸭在我国的多数地区都有分布。北京可见于郊区湿地或城市公园，整体数量不多，但遇见率较高，为旅鸟。

北京市重点保护 Beijing priority conservation	国家重点保护 State priority conservation	IUCN
是（YES）	——	无危（LC）2016.10.1

雌鸟 Female

A waterfowl with simple but elegant plumage, sexes similar. Breeding male mostly grayish; black barring on chest; black bill. Dull, brownish female similar to female Mallard. Difficult to tell apart from other ducks due to plain appearance. Most easily identified by bright white wing patch.

Inhabits inland aquatic environments such as lakes, reservoirs, and marshes. Prefers open water with rich aquatic plants. Form small flocks. Feeds mostly on aquatic plants.

Occurs in most regions in China. A passage migrant in Beijing where it is often found in suburban wetlands or urban parks. Easy to find despite a small overall population.

016 中文名 斑嘴鸭

学　名 *Anas zonorhyncha*

英文名 Chinese Spot-billed Duck

雁形目
ANSERIFORMES

鸭科
Anatidae

斑嘴鸭是体型较大的游禽，比一般家鸭略大。雌雄相似。斑嘴鸭全身羽毛以灰褐色为主，远观比较朴素，特点不够明显。头部具明显白色眉纹，喙黑色，前端具明显黄色。体羽多深褐色，翼镜具蓝绿色金属光泽。

斑嘴鸭常栖息于水库、湖泊和城市公园，偶尔会在海滨停息。越冬时喜成小群活动。斑嘴鸭善于游泳但不会潜水，主要以水草为食，也会捕食田螺、贝类、鱼虾等。

斑嘴鸭是我国比较常见的鸭科鸟类，广泛分布于各地。北京的郊区较为常见，为旅鸟和夏候鸟。

北京市重点保护 Beijing priority conservation	国家重点保护 State priority conservation	IUCN
——	——	无危（LC）2018.8.6

成鸟 Adult

A rather large waterfowl that is larger than usual domestic ducks, sexes similar. Plumage mostly grayish brown; dull looking from a far distance; no striking features; distinct white supercilium; dark bill with distinct yellow tip; body feathers mostly dark brown; speculum metallic blue and green.

Often inhabits reservoirs, lakes, and urban parks, but sometimes rests on beaches. Form small flocks in winter. Adept at swimming but does not dive. Feeds mostly on aquatic plants but can also prey on river snails, shellfish, fish, and shrimps.

A widely distributed common duck in China. A passage migrant and summer visitor in Beijing where it is common in the suburbs.

中文名 **绿头鸭**
学　名 *Anas platyrhynchos*
英文名 Mallard

雁形目
ANSERIFORMES
鸭科
Anatidae

绿头鸭是我国分布最广、数量最多、最被人熟知的游禽。雌雄差异大。雄鸟繁殖羽头部绿色，具金属光泽，在逆光情况下偏紫色。颈中部具一圈白色颈环，体色偏灰褐色，两枚特化的尾上覆羽向上卷曲。雌鸟整体灰褐色，喙橙红色。

我国本土的家鸭多数是从野生绿头鸭培育而来。它们喜欢在河流、湖泊等水域集群生活。主要以水生植物为食，有时捕捉鱼虾等水生动物。

绿头鸭广泛分布于我国各地。北京的郊区、城区公园常见，为旅鸟、冬候鸟和夏候鸟。

北京市重点保护 Beijing priority conservation	国家重点保护 State priority conservation	IUCN
——	——	无危（LC）2016.10.1

雌鸟 Female

Most widely distributed, abundant, and recognized waterfowl in China. Sexes different. Breeding male has metallic green head, which appears purplish under backlighting; white neck ring at the middle of its neck; plumage grayish brown; specialized uppertail-coverts upcurled. Female grayish brown overall with an orange bill.

Gregarious in aquatic environments such as rivers and lakes. Feeds mostly on aquatic plants but sometimes fish and shrimps as well. Most domestic ducks come from Mallards.

Widely distributed everywhere in China. A passage migrant, winter visitor, and summer visitor in Beijing where it is common in suburbs and urban parks.

018 中文名 绿翅鸭

学　名 *Anas crecca*

英文名 Eurasian Teal

雁形目 ANSERIFORMES

鸭科 Anatidae

绿翅鸭是一种小型游禽，体长不到40厘米。雌雄差异大。翼镜以具金属光泽的绿色为主，另沾黑色及白色。雄鸟繁殖羽羽色艳丽，头部偏栗色，眼周附近有一道明亮的绿色斑块。雌鸟的羽毛暗淡，以灰褐色为主。

绿翅鸭常栖于河流、湖泊等水域环境。冬季主要以植物嫩芽为食，其他季节除了植物也吃田螺、昆虫等小动物。它们多生活于湿地环境，有时至农田寻觅粮食。一般较为惧人。

绿翅鸭广泛分布于我国各地。北京可见于郊区、城市公园的各类水域环境，为旅鸟。

北京市重点保护 Beijing priority conservation	国家重点保护 State priority conservation	IUCN
—	—	无危（LC）2020.8.13

雌鸟 Female

A small waterfowl with body length smaller than 40cm; sexes different; speculum mostly green metallic, with black and white. Breeding male has colorful plumage; head chestnut-colored; green metallic patch around eye; female duller, mostly grayish brown.

Often inhabits aquatic environments such as rivers and lakes. Feeds mostly on buds of aquatic plants in winter but will also feed on river snails and insects in other seasons. Found in wetlands and sometimes feeding in farmland. Normally rather scared of people.

Widely distributed in China. A passage migrant in Beijing where it is seen in all kinds of aquatic environments in suburbs and urban parks.

019

中文名 **小䴙䴘**
学　名 *Tachybaptus ruficollis*
英文名 Little Grebe

䴙䴘目
PODICIPEDIFORMES
䴙䴘科
Podicipedidae

繁殖羽 Breeding

小䴙䴘，在野外常被错认成一种体型小的野鸭。雌雄相似。繁殖羽头顶深灰色，喙深灰色，喙角具一白斑，虹膜偏黄色，颈部深红色，体羽多深灰色。非繁殖羽整体多黄褐色。

小䴙䴘栖息于水流平缓的河流、湖泊以及沼泽中，尤好在长有芦苇等挺水植物的小池塘繁殖。脚上有蹼，善于游泳及潜水，以水中的小鱼、小虾为食。生性比较机警，如果遇到危险，会立刻钻入水中，潜行一段，再从水面其他处钻出。

小䴙䴘在我国大部分地区都有分布。北京的湖泊、河流、水库及诸多城市公园中常见，为留鸟、夏候鸟和旅鸟。

北京市重点保护 Beijing priority conservation	国家重点保护 State priority conservation	IUCN
是（YES）	—	无危（LC）2016.10.1

非繁殖羽 Non-breeding

Often misidentified as a small duck, sexes similar. Breeding plumage includes dark gray crown and bill, which has white spot at the tip; yellowish iris; dark red neck; body feathers mostly dark gray; non-breeding plumage mostly yellowish brown.

Inhabits slow-moving rivers, lakes, and marshes. Prefers to breed in ponds with emergent vegetation such as reed. Webbed feet. Adept at swimming and diving. Feeds on small fish and shrimps. Vigilant; in case of danger, dives into water, swims for a while, and comes back out somewhere else.

Occurs in most regions in China. A resident, summer visitor, and passage migrant in Beijing where it is common in lakes, rivers, reservoirs, and many urban parks.

中文名 **凤头䴙䴘**
学　名 *Podiceps cristatus*
英文名 Great Crested Grebe

䴙䴘目
PODICIPEDIFORMES

䴙䴘科
Podicipedidae

凤头䴙䴘在我国有分布的五种䴙䴘中是体型最大的一种，体长可达 50 厘米。雌雄相似。繁殖期头顶具一明显的灰褐色羽冠，脸颊白色及棕红色，白色颈部较长，背部深褐色。非繁殖羽体色暗淡。

凤头䴙䴘常栖于城市公园、湖泊、水库等地带，尤好生活在长有浓密芦苇、水草的湖里，以水中的鱼虾为食。每年春季的繁殖期，常见其在水面上跳独特的求偶舞蹈，鼓起羽冠，并肩踏水而行。

凤头䴙䴘在我国的多数地区都有分布。北京多见于郊区的水库，颐和园每年有繁殖个体，为夏候鸟和旅鸟。

北京市重点保护 Beijing priority conservation	国家重点保护 State priority conservation	IUCN
是（YES）	—	无危（LC）2019.8.11

非繁殖羽 Non-breeding

Largest among all five species in China, body length up to 50cm, sexes similar. Breeding adult has distinct grayish brown erectile crest; cheek white and rufous; neck white and long; back dark brown. Non-breeding plumage duller.

Often inhabits urban parks, lakes, and reservoirs. Particularly likes lakes with reeds and aquatic plants. Feeds on fish and shrimps. During the breeding season, pairs often seen performing unique mating display: running side-by-side on the water with erected crests.

Occurs in most regions in China. A summer visitor and passage migrant in Beijing where it is mostly found in suburban reservoirs; breeding individuals occur in the Summer Palace every year.

中文名 **普通鸬鹚**
学　名 *Phalacrocorax carbo*
英文名 Great Cormorant

鲣鸟目
SULIFORMES

鸬鹚科
Phalacrocoracidae

普通鸬鹚在民间被称为“鱼鹰”或者“黑鱼郎”，是一种大型游禽，体长可超过80厘米。雌雄相似。羽毛以黑色为主，喙尖利有钩似猛禽。头颈、肩、两翼有蓝色或紫色的金属光泽。繁殖羽头顶至颈侧沾白色。

普通鸬鹚常栖息于水库、河流环境，善于潜水，通常集群生活。有时集体下潜，在水中飞速游动追逐鱼群。不下水时，它们常会站在岸边的树枝、朽木上，有时还会伸开双翼，以晾干羽毛上的水。

普通鸬鹚广泛分布于我国各地。北京多见于水库环境，多为旅鸟。

北京市重点保护 Beijing priority conservation	国家重点保护 State priority conservation	IUCN
是（YES）	——	无危（LC）2018.8.9

A large waterfowl known by local people as "fish hawk" or "black fish dude". Body length can be over 80cm, sexes similar. Plumage mostly black; bill has hook like raptors; head, neck, shoulders, and wings have blue or purple metallic coloration; head and upper neck are white in breeding plumage.

Often inhabits reservoirs and rivers. Adept at diving. Usually form flocks, which sometimes dive together and rapidly chase after fish underwater. When on land, often perch on branches and snags. Sometimes spread out wings to dry.

Widely distributed everywhere in China. A passage migrant in Beijing where it is often seen in reservoirs.

中文名 **红嘴鸥**
学　名 *Chroicocephalus ridibundus*
英文名 Black-headed Gull

鸻形目
CHARADRIIFORMES
鸥科
Laridae

红嘴鸥属于中等体型的鸥类，体长约 40 厘米。雌雄相似。繁殖羽体羽多为白色，头部黑色，白色眼圈较窄，喙红色，翼尖黑色。非繁殖羽头部整体偏白色，耳羽具深色斑纹。亚成鸟喙橙色，体羽具黑色斑纹。

红嘴鸥适应能力极强，于海边或内陆湖泊、河流、水库等地带都可栖息。主要以鱼虾为食，通常成群活动。

红嘴鸥广泛分布于我国各地，数量多。北京郊区面积较大水域环境有机会看到，为旅鸟。

北京市重点保护 Beijing priority conservation	国家重点保护 State priority conservation	IUCN
——	——	无危（LC）2018.8.17

繁殖羽 Breeding

A medium-sized gull with body length roughly 40cm, sexes similar. Breeding plumage mostly white; black head; narrow, white eye-ring; red bill; black wingtips. Head of non-breeding plumage mostly whitish; dark spot on ear-coverts. Subadult has orange bill; body feathers with black markings.

Strong adaptability; inhabits coastal areas, inland lakes, rivers, and reservoirs. Feeds on fish and shrimps. Usually form flocks.

Abundant and widely distributed everywhere in China. A passage migrant in Beijing where it can be seen in large water bodies in suburban areas.

中文名 **普通燕鸥**
学　名 *Sterna hirundo*
英文名 Common Tern

鸻形目
CHARADRIIFORMES

鸥科
Laridae

普通燕鸥属于小型游禽，体长约 35 厘米。雌雄相似。翼窄长，尾长。成鸟头顶黑色，脸部白色，喙黑色或红色。上体灰色，下体白色，白色尾开叉较大。跗跖偏红色。亚成鸟上体具灰褐色斑纹，喙偏黄色。

普通燕鸥常栖息于沿海或内陆的多种湿地环境，擅在空中悬停，有时成群活动。主要以鱼虾为食。

普通燕鸥广泛分布于我国各地，数量多。北京郊区的沼泽湿地可见，为夏候鸟和旅鸟。

北京市重点保护 Beijing priority conservation	国家重点保护 State priority conservation	IUCN
——	——	无危（LC）2018.8.7

繁殖羽 Breeding

A small waterfowl with body length roughly 35cm, sexes similar. Narrow, long wings and long tails. Adult has black cap, white face, and black or red bill; gray upperparts and white underparts; white tail deeply forked; reddish tarsi. Subadult has grayish brown fringes on upperparts; yellowish bill.

Often inhabits diverse wetland environments in costal or inland areas. Adept at hovering in the air. Sometimes form flocks. Feeds mostly on fish and shrimps.

Abundant and widely distributed everywhere in China. A summer visitor and passage migrant in Beijing where it is found in suburban wetlands.

中文名 **黑水鸡**
学　名 *Gallinula chloropus*
英文名 Common Moorhen

鹤形目
GRUIFORMES
秧鸡科
Rallidae

黑水鸡是一种小型涉禽。雌雄相似。成鸟整体多为黑色，体侧及尾下覆羽白色，喙及额板红色，喙尖黄绿色，虹膜暗红色，跗跖黄绿色。幼鸟灰褐色。

黑水鸡栖息于平原地区的池塘、沼泽、水田等环境。性机警。多在晨昏活动，善游泳及在浮草上行走。主要以水生植物、昆虫、鱼虾为食。

黑水鸡在我国多数地区有分布。北京的城区及郊区均较为常见，多为留鸟和夏候鸟。

北京市重点保护 Beijing priority conservation	国家重点保护 State priority conservation	IUCN
——	——	无危（LC）2016.10.1

A small wading bird, sexes similar. Adult plumage mostly black; white flanks and undertail-coverts; red bill and frontal shield; bill tip greenish yellow; iris dark red; tarsi yellowish green. Juvenile grayish brown.

Inhabits environments such as ponds, marshes, and paddy fields in plains. Vigilant. Crepuscular. Adept at swimming and walking on floating vegetation. Feeds mostly on aquatic plants, insects, fish and shrimps.

Occurs in most regions in China. A resident and summer visitor in Beijing where it is common in urban and suburban areas.

025 中文名 白骨顶（骨顶鸡）

学　名 *Fulica atra*

英文名 Common Coot

鹤形目 GRUIFORMES

秧鸡科 Rallidae

白骨顶是一种小型涉禽，体型较黑水鸡大。雌雄相似。成鸟整体多为黑色，喙及额板白色，虹膜暗红色，跗跖灰绿色。

白骨顶栖息于湖泊、河流、沼泽、水田等环境。多在晨昏活动，善游泳及潜水。非繁殖期常成群活动。主要以水生植物、昆虫、鱼虾为食。

白骨顶在我国多数地区有分布。北京郊区的各类水域环境可见，亦见于颐和园、圆明园等城区公园，多为旅鸟。

保护级别 Protected-level	北京市重点保护 Beijing priority conservation	国家重点保护 State priority conservation	IUCN
	——	——	无危（LC）2019.8.14

成鸟 Adult

A small wading bird larger than a Common Moorhen, sexes similar. Adult mostly black; white bill and frontal shield; iris dark red; tarsi greenish gray.

Inhabits environments such as lakes, rivers, marshes, and paddy fields. Mostly crepuscular. Adept swimmer and diver. Form flocks in the non-breeding season. Feeds mostly on aquatic plants, insects, fish, and shrimps.

Occurs in most regions in China. Populations in Beijing are mostly migratory; found in all kinds of aquatic environments, also seen in urban parks such as the Summer Palace and Yuanmingyuan.

中文名 **灰鹤**
学　名 *Grus grus*
英文名 Common Crane

鹤形目
GRUIFORMES

鹤科
Gruidae

灰鹤是一种整体为灰色的大型涉禽。雌雄相似。成鸟头顶红色，颈部、眼先、下颏黑色，颈侧及体羽多灰色，两翼及尾沾黑色。

灰鹤栖息于开阔的农田、草地等环境。性机警惧人。非繁殖期常成大群活动。主要以植物为食。

灰鹤在我国多数地区有分布。北京郊区的延庆、密云等地可见，多为旅鸟和冬候鸟。

北京市重点保护 Beijing priority conservation	国家重点保护 State priority conservation	IUCN
——	二级（II）	无危（LC）2016.10.1

成鸟 Adult

A large, mostly gray wading bird, sexes similar. Adult has red crown patch; black neck, lore, and chin; side of neck and body feathers mostly gray; black on wings and tail.

Inhabits environments such as open agricultural fields and grasslands. Vigilant. Form large flocks in the non-breeding season. Feeds mostly on plant material.

Occurs in most regions in China. A passage migrant and winter visitor in Beijing where it is found in places such as Yanqing and Miyun.

027

中文名 **黑鹳**
学　名 *Ciconia nigra*
英文名 Black Stork

鹳形目
CICONIIFORMES

鹳科
Ciconiidae

黑鹳是一种大型涉禽。雌雄相似。成鸟头部、颈部、胸部、背部具黑色及绿色金属色光泽，眼周具红色裸皮，喙红色，腹部白色，跗跖红色。亚成鸟体羽多灰色，少金属光泽。

黑鹳栖息于河流、淡水湖泊、沼泽等湿地环境。性机警，较惧人。有时成小群活动。主要以鱼类为食，有时亦捕捉两栖动物、甲壳类等。

黑鹳在我国除西藏外多数地区有分布。北京郊区的十渡较为常见，多为留鸟。

北京市重点保护 Beijing priority conservation	国家重点保护 State priority conservation	IUCN
—	一级（I）	无危（LC）2016.10.1

亚成鸟 Immature

A large wading bird, sexes similar. Adult has black and green metallic coloration on the head, neck, chest, and back; red bare parts around eye; red bill; white belly, red tarsi. Subadult mostly gray; less metallic coloration.

Inhabits wetland environments such as rivers, freshwater lakes, and marshes. Vigilant and afraid of people. Sometimes form small flocks. Feeds mostly on fish, but sometimes amphibians and crustaceans as well.

Occurs in most of China except for Xizang. A resident in Beijing where it is common in Shidu.

中文名 **黄斑苇鳽（黄苇鳽）**
学　名 *Ixobrychus sinensis*
英文名 Yellow Bittern

鹈形目
PELECANIFORMES
鹭科
Ardeidae

雄鸟 Male

黄斑苇鳽是一种小型涉禽。雌雄相似。成鸟头顶黑色，头部黄色，上喙偏黑色，下喙黄色。上体黄褐色，下体皮黄色，喉部至胸腹部具黄色纵纹，飞羽及尾黑色。雌鸟似雄鸟，但头顶黑色较浅，下体纵纹较明显。幼鸟体羽斑驳。

黄斑苇鳽栖息于平原、丘陵地带的沼泽、池塘、稻田等各类湿地环境。性隐秘，常单独或成对活动，常潜藏在芦苇丛中，有时伸长颈部将喙向上举，拟态周边环境。主要以鱼虾、两栖动物为食。

黄斑苇鳽分布于我国东部和中部多数地区。北京郊区及城区的湿地环境可见，为夏候鸟和旅鸟。

北京市重点保护 Beijing priority conservation	国家重点保护 State priority conservation	IUCN
—	—	无危（LC）2016.10.1

A small wading bird, sexes similar. Adult has black cap, yellow head, blackish maxilla, and yellow lower mandible; upperparts yellowish brown while underparts lighter-colored; yellow streaks from throat to chest and belly; black tail and flight feathers. Females similar but has paler cap; underparts with distinct streaking. Juvenile body feathers covered with spots.

Inhabits wetland environments such as marshes, ponds, and paddy fields in plains and hills. Secretive; usually solitary or in pairs; often hidden in dense reed; sometimes stretches out its neck upwards, thereby mimicking the surrounding environment. Feeds mostly on fish, shrimps, and amphibians.

Occurs in most regions in eastern and central China. A summer visitor and passage migrant in Beijing where it is found in urban and suburban wetland environments.

中文名 **夜鹭**
学　名 *Nycticorax nycticorax*
英文名 Black-crowned Night-heron

鹈形目
PELECANIFORMES
鹭科
Ardeidae

夜鹭是一种中型涉禽。雌雄相似。成鸟头顶蓝黑色，头部具一明显的白色羽辫，喙黑色，虹膜红色。上体蓝黑色，下体白色及灰色。跗跖黄绿色。幼鸟棕褐色，体羽密布白色斑点。

夜鹭栖息于河流、淡水湖泊、沼泽、池塘、红树林等各类湿地环境。有时集小群活动。繁殖期集群营巢于树枝上。主要以鱼类、两栖动物为食。

夜鹭分布于我国多数地区。北京郊区及城区的湿地环境可见，多为夏候鸟和旅鸟。

北京市重点保护 Beijing priority conservation	国家重点保护 State priority conservation	IUCN
—	—	无危（LC）2016.10.1

亚成鸟 Immature

A medium-sized wading bird, sexes similar. Adult has bluish black cap, white head plumes, black bill, red iris, bluish black upperparts, white and gray underparts, and greenish yellow tarsi. Juvenile brownish and covered with white spots.

Inhabits wetland environments such as rivers, freshwater lakes, marshes, ponds, and mangroves. Sometimes form small flocks. Form breeding colonies on tree branches in the breeding season. Feeds mostly on fish and amphibians.

Occurs in most regions in China. A summer visitor and passage migrant in Beijing where it is found in urban and suburban wetland environments.

030 中文名 池鹭

学　名 *Ardeola bacchus*

英文名 Chinese Pond Heron

鹈形目 PELECANIFORMES

鹭科 Ardeidae

池鹭是一种中型涉禽。雌雄相似。成鸟繁殖羽头部棕红色，喙黄绿色，喙尖黑色。胸部红色，下腹部白色，背部蓝灰色。跗跖橙色。非繁殖羽羽色较淡，头部多棕褐色纵纹。幼鸟似成鸟非繁殖羽，但体羽更为斑驳。

池鹭栖息于淡水湖泊、沼泽、池塘等各类湿地环境。常单独或成对活动。主要以鱼虾、两栖动物、昆虫等为食。

池鹭分布于我国东部及中部多数地区。北京郊区及城区的湿地环境可见，多为夏候鸟和旅鸟。

北京市重点保护 Beijing priority conservation	国家重点保护 State priority conservation	IUCN
——	——	无危（LC）2016.10.1

A medium-sized wading bird, sexes similar. Breeding adult has chestnut head and chest, greenish yellow bill with black tip, white lower belly, bluish gray back, and orange tarsi. Duller non-breeding plumage has brown streaking on head. Juvenile similar to non-breeding adult, but with spotted body feathers.

Inhabits wetland environments such as freshwater lakes, marshes, and ponds. Often feeds solitarily or in pairs. Feeds mostly on fish, shrimp, amphibians, and insects.

Occurs in most regions in eastern and central China. A summer visitor and passage migrant in Beijing where it is found in urban and suburban wetland environments.

中文名 **苍鹭**
学　名 *Ardea cinerea*
英文名 Grey Heron

鹈形目
PELECANIFORMES
鹭科
Ardeidae

苍鹭是整体以灰色为主的大型涉禽。雌雄相似。成鸟头部灰色，头部具一蓝黑色羽辫，颈部甚长，为灰白色，具蓝黑色纵纹，体羽多灰色。成鸟繁殖羽喙橙红色，非繁殖羽多为黄色。

苍鹭栖息于河流、淡水湖泊、沼泽等各类湿地环境。非繁殖期常成群活动。主要以鱼类、两栖动物、昆虫等为食，有时亦捕捉小型鸟类及小型兽类。

苍鹭分布于我国多数地区。北京郊区及城区的湿地环境可见，多为夏候鸟、旅鸟和留鸟。

北京市重点保护 Beijing priority conservation	国家重点保护 State priority conservation	IUCN
—	—	无危（LC）2019.8.13

成鸟 Adult

A large, mostly gray wading bird, sexes similar. Adult has gray head with bluish black crest plumes; long, grayish white neck with bluish black streaking; body feathers mostly gray. Breeding adult has orange bill; non-breeding plumage mostly yellow.

Inhabits wetland environments such as rivers, freshwater lakes, and marshes. Form flocks in the non-breeding season. Feeds on fish, amphibians, and insects, but sometimes on smaller birds and mammals as well.

Occurs in most regions in China. A summer visitor, passage migrant, and resident in Beijing where it is found in urban and suburban wetland environments.

中文名 **大白鹭**
学　名 *Ardea alba*
英文名 Great Egret

鹈形目
PELECANIFORMES
鹭科
Ardeidae

大白鹭是整体以白色为主的大型涉禽。雌雄相似。成鸟周身白色，繁殖羽喙黑色，眼先蓝绿色；非繁殖羽喙黄色。

大白鹭栖息于河流、淡水湖泊、沼泽、红树林等各类湿地环境。常成小群活动。主要以鱼类、两栖动物、昆虫等为食，有时亦捕捉小型鸟类及小型兽类。

大白鹭在我国多数地区有分布。北京郊区及城区的湿地环境可见，多为夏候鸟和旅鸟。

北京市重点保护 Beijing priority conservation	国家重点保护 State priority conservation	IUCN
是（YES）	——	无危（LC）2016.10.1

A large and mostly white wading bird, sexes similar. Adult all white. Breeding plumage has black bill and bluish green lore. Non-breeding plumage has yellow bill.

Inhabits wetland environments such as rivers, freshwater lakes, and marshes. Often form small flocks. Feeds on fish, shrimps, amphibians, and insects, but sometimes on smaller birds and mammals as well.

Occurs in most regions in China. Mostly a summer visitor and passage migrant in Beijing where it can be seen in urban and suburban wetland environments.

033 中文名 白鹭

学　名 *Egretta garzetta*
英文名 Little Egret

鹈形目
PELECANIFORMES
鹭科
Ardeidae

白鹭是一种整体为白色的中型涉禽。雌雄相似。成鸟周身白色，喙黑色，跗跖黑色，趾黄色。繁殖羽头后具一白色羽辫。

白鹭栖息于河流、淡水湖泊、沼泽、水田、红树林、海岸边等各类湿地环境。有时成小群活动，觅食时常用脚搅动水底，惊出底栖动物后捕食。主要以鱼虾、两栖动物、昆虫等为食。

白鹭分布于我国大多数地区。北京郊区及城区的湿地环境可见，多为夏候鸟和旅鸟。

保护级别 Protected-level	北京市重点保护 Beijing priority conservation	国家重点保护 State priority conservation	IUCN
	——	——	无危（LC）2016.10.1

繁殖羽 Breeding

A medium-sized white wading bird, sexes similar. Adult all white; black bill and tarsi; yellow toes. Breeding plumage has white ornamental plumes.

Inhabits wetland environments such as rivers, freshwater lakes, marshes, paddy fields, mangroves, and coastal areas. Sometimes form small flocks. Often forage using foot stirring, which scares up benthic organisms in the water. Feeds mostly on fish, shrimps, amphibians, and insects.

Occurs in most regions in China. Mostly a summer visitor and passage migrant in Beijing where it occurs in urban and suburban wetland environments.

中文名 **黑翅长脚鹬**
学　名 *Himantopus himantopus*
英文名 Black-winged Stilt

鸻形目
CHARADRIIFORMES

反嘴鹬科
Recurvirostridae

雄鸟 Male

黑翅长脚鹬是非常容易识别的鸻鹬类。雌雄相似。上体黑色，下体白色，黑色喙细长似针，红色腿甚长。雄鸟背部偏黑色，雌鸟偏褐色。亚成鸟喙及腿色浅，背部偏黄褐色。

黑翅长脚鹬常栖息于沼泽湿地和盐田等地带。时常成小群在浅水滩、沼泽寻找泥中的鱼虾及昆虫等食物。

黑翅长脚鹬分布于我国多数地区。北京延庆等地的淡水湿地环境不难见到，为夏候鸟和旅鸟。

北京市重点保护 Beijing priority conservation	国家重点保护 State priority conservation	IUCN
——	——	无危（LC）2016.10.1

An easily identifiable shorebird, sexes similar. Black upperparts and white underparts; black bill slender and needle-like; Long, pinkish legs; back black on male and brownish on female. Subadult has lighter-colored bill and legs, and yellowish-brown back.

Inhabits marshes and salt marshes. Often form small flocks that feed on fish, shrimps, and insects in shallow water.

Occurs in most regions in China. A summer visitor and passage migrant in Beijing where it is often seen in freshwater wetlands in places such as Yanqing.

中文名 **凤头麦鸡**
学　名 *Vanellus vanellus*
英文名 Northern Lapwing

鸻形目
CHARADRIIFORMES
鸻科
Charadriidae

凤头麦鸡是一种体长约 30 厘米的小型涉禽。雌雄相似。阴天看整体黑白两色。晴天观察，其羽色非常艳丽。上体金属铜绿色沾有紫红色，头上具一显著的向上竖起的羽冠，胸部近黑色，腹部偏白。亚成鸟体色较为暗淡。

凤头麦鸡喜欢在沼泽地和湿润草地、农田等地带活动。善飞行，常在空中上下翻飞，尤其是天敌接近其巢区时，会尖声报警并奋力驱逐。主要以昆虫为食。

凤头麦鸡广布于我国多地。北京的郊区不难遇到，为较常见旅鸟。

保护级别 Protected-level	北京市重点保护 Beijing priority conservation	国家重点保护 State priority conservation	IUCN
	——	——	近危（NT）2016.10.1

成鸟 Adult

A shorebird with body length about 30cm, sexes similar. Appears black and white on cloudy days, but more colorful on sunny days; upperparts mostly bronze and green but with some purple; distinct, erected crest; black chest and white belly. Subadult duller.

Likes marshes, moist grasslands, and farm fields. Adept at flying; often seen flying up and down in the air, especially when predators approach its nesting area; will sound off high-pitched alarms and mob predators. Feeds mostly on insects.

Occurs in many regions in China. A common passage migrant in Beijing where it can be easily spotted in suburban areas.

中文名 金眶鸻

学　名 *Charadrius dubius*

英文名 Little Ringed Plover

鸻形目
CHARADRIIFORMES

鸻科
Charadriidae

金眶鸻是一种小型涉禽，体长仅 15 厘米左右。雌雄相似。整体为灰褐色，额头白色，一道黑色斑纹自头顶延伸至眼后，具一明显的金黄色眼圈。喉部到下腹为白色，具一条明显的黑色胸带。亚成鸟体色暗淡。

金眶鸻栖息于河流、沼泽、草地等环境，常成对或成小群活动。它们喜在水岸边快速奔跑，寻找草籽、小昆虫为食。

金眶鸻在我国分布较为广泛，几乎所有省份都有记录。北京郊区的沼泽湿地可见，多为夏候鸟。

北京市重点保护 Beijing priority conservation	国家重点保护 State priority conservation	IUCN
是（YES）	—	无危（LC）2016.10.1

非繁殖羽 Non-breeding

A small shorebird only about 15cm, sexes similar. Grayish brown overall; white forehead; black stripe stretches from crown to auriculars; distinct golden-yellow eye-ring; all white from throat to lower belly; distinct black chest band. Subadult duller.

Inhabits rivers, marshes, grasslands. Form pairs or small flocks. Likes running on the shore. Searches for grass seeds and small insects.

Widely distributed in almost every province and city. A summer visitor in Beijing where it is found in suburban marshes.

中文名 **环颈鸻**
学　名 *Charadrius alexandrinus*
英文名 Kentish Plover

鸻形目
CHARADRIIFORMES
鸻科
Charadriidae

环颈鸻大小与金眶鸻相当。雌雄相似。雄鸟繁殖羽头顶棕色，贯眼纹黑色，颈侧有一道较窄的黑色斑纹，上体灰褐色，下体白色。雌鸟似雄鸟，但头顶颜色较淡。非繁殖羽及亚成鸟体色偏淡褐色。

环颈鸻栖息于沿海滩涂、盐田、内陆河流、草地、农田等环境，性格活泼好动，善于在地面快速奔走。主要以昆虫、软体动物为食。东部沿海地区迁徙时可见百只以上的大群。

环颈鸻分布于我国多数地区。北京郊区的湿地环境可见，为夏候鸟和旅鸟。

北京市重点保护 Beijing priority conservation	国家重点保护 State priority conservation	IUCN
——	——	无危（LC）2016.10.1

雌鸟 Female

Same size as the Little Ringed Plover, sexes similar. Breeding adult has brownish crown, black eye-stripe, and black patches at sides of chest; grayish brown upperparts and white underparts. Female similar to male, but with duller crown. Subadult and non-breeding individual mostly lighter brown.

Inhabits environments such as coastal beaches, salt marshes, inland rivers, grasslands, and agricultural fields. Very active and adept at running fast on the ground. Feeds mostly on insects and mollusks. During migration, flocks of more than 100 individuals can be seen in coast areas in eastern China.

Occurs in most regions in China. A summer visitor and passage migrant in Beijing where it is found in suburban wetlands.

038

中文名 **普通雨燕**
学　名 *Apus apus*
英文名 Common Swift

雨燕目
APODIFORMES

雨燕科
Apodidae

普通雨燕是一种小型攀禽。雌雄相似。体羽多为灰褐色，喉部白色，两翼狭长，尾开叉较为明显。

普通雨燕喜栖息于城区古建筑群环境。飞行速度甚快，转弯灵活，并常伴随高声鸣叫。喜成群活动。主要以昆虫为食。

普通雨燕在我国分布于东北、华北、西北等地区。北京城区的古建筑群环境较为常见，多为夏候鸟。

北京市重点保护 Beijing priority conservation	国家重点保护 State priority conservation	IUCN
是（YES）	——	无危（LC）2016.10.1

成鸟 Adult

A small swift, sexes similar. Body feathers mostly grayish brown; white throat; long, narrow wings; tail distinctly forked.

Likes inhabiting old building complexes in urban areas. Flies rather fast and turns agilely; often gives high-pitched flight calls. Form flocks. Feeds mostly on insects.

Occurs in northwestern, north, and northeastern China. A signature summer visitor in Beijing where it is common in old building complexes.

中文名 **四声杜鹃**
学　名 *Cuculus micropterus*
英文名 Indian Cuckoo

鹃形目
CUCULIFORMES
杜鹃科
Cuculidae

四声杜鹃是一种中型攀禽。雌雄相似。头部灰色或灰褐色，眼圈黄色，虹膜暗褐色，喙黑色，下喙基黄色，胸腹部偏白色，具深色横纹，背部、两翼及尾偏灰色。

四声杜鹃喜栖息于平原或低海拔地区的林地环境。巢寄生鸟类，在北京地区喜将卵寄生于灰喜鹊巢中。常单独活动。主要以昆虫为食。

四声杜鹃分布于我国除西北地区外的多数地区。北京城区及低海拔的郊区较为常见，多为夏候鸟和旅鸟。

北京市重点保护 Beijing priority conservation	国家重点保护 State priority conservation	IUCN
是（YES）	—	无危（LC）2018.8.6

成鸟 Adult

A medium-sized cuckoo, sexes similar. head gray or grayish brown; yellow eye-ring and brown iris; black bill with the base of lower mandible yellow; underparts mostly white with black bands; back, wings, and tails mostly gray.

Inhabits woodland environments in plains or low-elevation regions. Brood parasite. In Beijing, often lays eggs in nests of Azure-winged Magpies. Usually solitary. Feeds mostly on insects.

Occurs in most regions in China except northwestern China. A summer visitor and passage migrant in Beijing where it is common in cities and suburban areas at lower elevations.

中文名 **大杜鹃**
学　名 *Cuculus canorus*
英文名 Common Cuckoo

鹃形目
CUCULIFORMES
杜鹃科
Cuculidae

大杜鹃是一种中型攀禽。雌雄相似。头部灰色，眼圈黄色，虹膜黄色，喙黑色，下喙基黄色；胸腹部偏白色，具深色横纹，横纹较四声杜鹃浅；背部、两翼及尾偏灰色。

大杜鹃适应能力甚强，可栖息于平原至高海拔地区的湿地、灌丛、林地等多种环境。巢寄生鸟类，在北京地区喜将卵寄生于东方大苇莺巢中。常单独或成对活动。主要以昆虫为食。

大杜鹃分布于我国多数地区。北京城区及低海拔的郊区湿地环境较为常见，多为夏候鸟和旅鸟。

北京市重点保护 Beijing priority conservation	国家重点保护 State priority conservation	IUCN
是（YES）	——	无危（LC）2016.10.1

成鸟 Adult

A medium-sized cuckoo, sexes similar. Gray head; yellow eye-ring and iris; black bill with the base of the lower mandible yellow; underparts mostly white with black bands, which are less pronounced than those of the Indian Cuckoo; back, wings, and tails mostly gray.

Adapts to a variety of environments such as wetlands, shrubs, and woodlands in plains and at high elevations. Brood parasite; in Beijing often lays eggs in nests of Oriental Reed Warblers. Usually solitary or in pairs. Feeds mostly on insects.

Occurs in most regions in China. Mostly a summer visitor and passage migrant in Beijing where it is often found in wetland environments in cities and suburban areas at lower elevations.

中文名 **戴胜**
学　名 *Upupa epops*
英文名 Eurasian Hoopoe

犀鸟目
BUCEROTIFORMES

戴胜科
Upupidae

戴胜是一种中型攀禽。雌雄相似。头部黄褐色，头顶具明显的黄褐色、黑色、白色羽冠，黑色喙长而下弯，下喙基褐色，上体及尾羽多为黑白两色，下体黄褐色。

戴胜适应能力强，可栖息于各海拔高度的开阔地、草地、公园、果园、稀疏林地等多种环境。常单独或成对活动。主要以昆虫为食。

戴胜分布于我国多数地区。北京城区及郊区均可见，多为夏候鸟和旅鸟。

北京市重点保护 Beijing priority conservation	国家重点保护 State priority conservation	IUCN
是（YES）	——	无危（LC）2020.8.14

成鸟 Adult

A medium-sized scansorial bird, sexes similar. Head yellowish brown; distinct crest consists of yellowish brown, black, and white; dark, long, and downcurved bill with the base of the lower mandible brown; upperparts and tail feathers mostly black and white; underparts yellowish brown.

High adaptability; inhabits diverse environments such as open habitats, grasslands, parks, orchards, and sparse woodlands in all elevational ranges. Usually solitary or in pairs. Feeds mostly on insects.

Occurs in most regions in China. A summer visitor and passage migrant in Beijing where it is found in urban and suburban areas.

中文名 **普通翠鸟**
学　名 *Alcedo atthis*
英文名 Common Kingfisher

佛法僧目
CORACIIFORMES

翠鸟科
Alcedinidae

普通翠鸟是一种小型攀禽。雌雄相似。头部、脸颊、两翼翠绿色，具白色点状斑纹，下颏及颈侧白色，胸腹部橙红色，背部亮蓝色，翠绿色尾羽甚短。雄鸟喙黑色，雌鸟上喙黑色，下喙橙红色。

普通翠鸟常栖息于低海拔的河流、池塘、湖泊等各类水域环境。常单独或成对活动。喜站立于水域上方的树枝上，发现鱼后，快速扎入水中捕食。主要以小型鱼类为食。

普通翠鸟分布于我国多数地区。北京城区及郊区均可见，多为留鸟和夏候鸟。

北京市重点保护 Beijing priority conservation	国家重点保护 State priority conservation	IUCN
是（YES）	——	无危（LC）2016.10.1

A small kingfisher, sexes similar. Crown, malar stripe, and wings bluish green with white spots; chin and sides of neck white; chest and belly rufous; bright blue back; bluish green tail rather short; bill all black in male while black (upper) and orange-red (lower) in female.

Often inhabits aquatic environments such as rivers, ponds, and lakes at low elevations. Usually solitary or in pairs. Hunts by perching on a branch above the water; upon detecting fish, quickly dives into the water; feeds mostly on small fish.

Occurs in most regions in China. A resident and summer visitor in Beijing where can be seen in urban and suburban areas.

中文名 **灰头绿啄木鸟**
学　名 *Picus canus*
英文名 Grey-faced Woodpecker

啄木鸟目
PICIFORMES

啄木鸟科
Picidae

灰头绿啄木鸟是一种中型攀禽。雌雄相似。头、颈灰色，雄鸟头顶红色，雌鸟头部无红色；下体灰绿色；背、腰、尾上覆羽、两翼覆羽灰绿色。

灰头绿啄木鸟常栖息于平原、中低海拔山区的城市公园、果园、阔叶林、混交林等环境，营巢于树洞。常单独或成对活动。主要以昆虫和植物果实为食。

灰头绿啄木鸟分布于我国多数地区。北京城区及郊区常见，为留鸟。

北京市重点保护 Beijing priority conservation	国家重点保护 State priority conservation	IUCN
是（YES）	—	无危（LC）2016.10.1

A medium-sized greenish woodpecker, sexes similar. Grey head and neck, male has red crown while female has no red on head; greyish-green underparts; greyish-green on back, rump, uppertail-coverts and wing-coverts.

Often inhabits environments such as urban parks, orchards, broadleaf forests, and mixed forests in plains and mountainous areas at low and middle elevations. Nests in tree cavities. Usually solitary or in pairs. Feeds mostly on insects and fruits.

Occurs in most regions in China. A resident in Beijing where it is common in urban and suburban areas.

中文名 **星头啄木鸟**
学　名 *Picoides canicapillus*
英文名 Grey-capped Woodpecker

啄木鸟目
PICIFORMES
啄木鸟科
Picidae

星头啄木鸟是一种小型攀禽。雌雄相似。是北京地区常见啄木鸟中体型最小的一种。头部主要为灰褐色，头顶黑色；上体黑色，具明显白斑；下体棕褐色，具黑色纵纹；尾黑色。

星头啄木鸟常栖息于平原和低海拔地区的城市公园、果园、针叶林、混交林等林地环境，营巢于树洞。常单独或成对活动。主要以昆虫为食。

星头啄木鸟分布于我国多数地区。北京城区及郊区常见，为留鸟。

北京市重点保护 Beijing priority conservation	国家重点保护 State priority conservation	IUCN
是（YES）	—	无危（LC）2016.10.1

成鸟 Adult

A small woodpecker, sexes similar. Smallest woodpecker among all common ones in Beijing. Grayish brown head; black crown; upperparts black with distinct white patches; underparts brownish with black streaks; black tail.

Often inhabits woodland environments such as urban parks, orchards, broadleaf forests, and mixed forests in plains and mountainous areas at low elevations. Nests in tree cavities. Usually solitary or in pairs. Feeds mostly on insects.

Occurs in most regions in China. A resident in Beijing where it is common in urban and suburban areas.

中文名 **大斑啄木鸟**
学　名 *Dendrocopos major*
英文名 Great Spotted Woodpecker

啄木鸟目
PICIFORMES

啄木鸟科
Picidae

大斑啄木鸟是一种中型攀禽。雌雄相似。颊部黄褐色，头顶黑色，颈侧具黑色“X”形条纹；上体黑色，具明显白斑；下体黄褐色，无纵纹；尾下覆羽红色，尾黑色。雄鸟枕部红色，雌鸟为黑色。

大斑啄木鸟常栖息于平原、丘陵、山地的阔叶林、混交林及果园等林地环境，营巢于树洞。常单独或成对活动。主要以昆虫为食。

大斑啄木鸟分布于我国多数地区。北京城区及郊区常见，为留鸟。

北京市重点保护 Beijing priority conservation	国家重点保护 State priority conservation	IUCN
是（YES）	——	无危（LC）2016.10.1

A medium-sized woodpecker, sexes similar. Cheek yellowish brown and crown black; black “X” on neck; upperparts black with distinct white patches; underparts yellowish brown without streaking; red undertail-coverts and black tail; nape red on male black on female.

Inhabits woodland environments such as broadleaf forests, mixed forests, and orchards in plains, hills, and mountains. Usually solitary or in pairs. Feeds mostly on insects.

Occurs in most regions in China. A resident in Beijing where it is common in urban and suburban areas.

中文名 **红角鸮**
学　名 *Otus sunia*
英文名 Oriental Scops Owl

鸮形目
STRIGIFORMES

鸱鸮科
Strigidae

红角鸮是一种小型猛禽。雌雄相似。整体多棕褐色，虹膜黄色，喙黑色，头顶两侧具角状耳羽；上体棕褐色，具白斑；下体黄褐色，具黑色纵纹。

红角鸮常栖息于山地和平原地区的阔叶林和混交林环境，有时亦见于城市公园和居民点。营巢于天然树洞中或啄木鸟废弃的旧洞，偶尔会在人工巢箱繁殖。夜行性，主要以昆虫和鼠类为食。

红角鸮分布于我国东部多数地区。北京郊区山地及城区部分公园可见，为旅鸟和夏候鸟。

	北京市重点保护 Beijing priority conservation	国家重点保护 State priority conservation	IUCN
	——	二级（II）	无危（LC）2016.10.1

成鸟 Adult

A small raptor, sexes similar. Mostly brown; yellow iris; black bill; horn-shaped ear-tufts; upperparts brown with white patches; underparts yellowish brown with black streaking.

Inhabits broadleaf and mixed forests in plains and mountainous areas, sometimes also found in urban parks and residential areas. Nests in natural tree cavities or ones made by woodpeckers; sometimes nests in artificial nest boxes. Nocturnal. Feeds mostly on insects and rodents.

Occurs in most regions in eastern China. A passage migrant and summer visitor in Beijing where it can be seen in urban parks and mountainous area.

中文名 **雀鹰**
学　名 *Accipiter nisus*
英文名 Eurasian Sparrow Hawk

鹰形目
ACCIPITRIFORMES
鹰科
Accipitridae

雀鹰是一种中型猛禽。雌雄相似。雄鸟头部灰色，脸颊红色，虹膜橙红色；上体青灰色；下体偏白色，具红褐色横纹。雌鸟头部褐色，眉纹白色；上体褐色；下体偏白色，具褐色横纹。亚成鸟似雌鸟，但腹部斑纹多为三角状。

雀鹰常栖息于平原、丘陵、山地的各类林地环境。常单独活动。性凶猛，会驱逐进入领地的其他猛禽。主要以中小型鸟类为食。

雀鹰分布于我国多数地区。北京城区公园及山区常见，为旅鸟和冬候鸟。

保护级别 Protected-level	北京市重点保护 Beijing priority conservation	国家重点保护 State priority conservation	IUCN
	—	二级（II）	无危（LC）2016.10.1

A medium-sized raptor, sexes similar. Male has gray head, reddish cheek, orange-red iris, slaty-gray upperparts, and whitish underparts with reddish brown barring. Female has brown head, white supercilium, brown upperparts, and whitish underparts with brown barring. Subadult similar to female, but underparts with triangle-shaped barring.

Inhabits woodland environments in plains, hills, and mountains. Usually solitary. Aggressive; will chase off other territorial intruders. Feeds mostly on small to medium-sized birds.

Occurs in most regions in China. A passage migrant and winter visitor in Beijing where it is common in urban parks and mountainous areas.

中文名 **白腹鹞**
学　名 *Circus spilonotus*
英文名 Eastern Marsh Harrier

鹰形目
ACCIPITRIFORMES
鹰科
Accipitridae

白腹鹞是一种中型猛禽。雌雄差异大。大陆型雄鸟头黑色或灰色，腹部、翼下白色。大陆型雌鸟整体颜色浅，腹部纵纹明显。日本型雌鸟或幼鸟深褐色，腹部颜色深，初级飞羽白色，有少量点斑。

白腹鹞常栖息于沼泽地、江河、湖泊、苇塘等各类湿地水域环境。喜单独或成对活动。多在湿地上空贴近水面或沼泽地低空飞行。主要以小型鸟类、蛙类、蛇类、鼠类为食。

白腹鹞分布于我国多数地区。北京延庆、密云等郊区较为常见，多为旅鸟。

北京市重点保护 Beijing priority conservation	国家重点保护 State priority conservation	IUCN
——	二级（II）	无危（LC）2016.10.1

日本型 / 亚成鸟 Immature

A medium-sized raptor, sexes different. Male of Chinese population has black or gray head, white belly, and mostly white underwing; female lighter-colored with distinct streaking on belly. Female and juvenile of Japanese population dark brown with dark belly; bright primary feathers; no streaking.

Inhabits wetland environments such as marshes, rivers, lakes, and reedbeds. Usually solitary or in pairs. Often flies low to the water in wetlands and marshes. Feeds mostly on small birds, frogs, snakes, and rodents.

Occurs in most regions in China. A passage migrant in Beijing where it is common in suburbs such as Yanqing and Miyun.

中文名 白尾鹞

学　名 *Circus cyaneus*

英文名 Hen Harrier

鹰形目
ACCIPITRIFORMES

鹰科
Accipitridae

白尾鹞是一种中型猛禽。雌雄差异大。雄鸟头灰色，腹部较白，翼下及背灰色，初级飞羽黑色。雌鸟深褐色，腹部纵纹重，翼下具明显横纹，腰部白色。成鸟虹膜黄色。亚成鸟似雌鸟，但虹膜色深。

白尾鹞常栖息于湖泊、河流、苇塘等各类湿地环境。喜单独或成对活动。多在湿地上空贴近水面或沼泽地低空飞行。主要以小型鸟类、蛙类、蛇类、鼠类为食。

白尾鹞分布于我国多数地区。北京延庆、密云、怀柔等郊区较为常见，为旅鸟和冬候鸟。

北京市重点保护 Beijing priority conservation	国家重点保护 State priority conservation	IUCN
——	二级（II）	无危（LC）2016.10.1

A medium-sized raptor, sexes different. Male has gray head, white belly, gray back and underwing, and black outer primaries. Female dark brown with heavy streaking on the belly, distinct underwing barring, and white rump. Adult has yellow iris. Subadult similar to female, but with darker iris.

Inhabits wetland environments such as lakes, rivers, and reedbeds. Usually solitary or in pairs. Often flies low to the water in wetlands and marshes. Feeds mostly on small birds, frogs, snakes, and rodents.

Occurs in most regions in China. A passage migrant and winter visitor in Beijing where it is common in suburbs such as Yanqing, Miyun, and Huairou.

050

中文名 **普通鵟**

学　名 *Buteo japonicus*

英文名 Eastern Buzzard

鹰形目
ACCIPITRIFORMES

鹰科
Accipitridae

普通鵟是一种中型猛禽。雌雄相似。北京所见个体一般全身黄褐色为主，两翼较宽，头部圆润，尾较短，翼下具黑色“腕斑”。亚成鸟腹部纵纹明显。

普通鵟常栖息于山地森林和山脚平原与草原地区，冬季常至旷野农田、荒地、村庄等地带活动。常单独或成对活动，迁徙时成群活动。主要以鼠类、蛙类、中小型鸟类为食。

普通鵟分布于我国东部多数地区。北京延庆、密云、怀柔等郊区较为常见，为旅鸟和冬候鸟。

北京市重点保护 Beijing priority conservation	国家重点保护 State priority conservation	IUCN
——	二级（II）	无危（LC）2016.10.1

成鸟 Adult

A medium-sized raptor, sexes similar. Individuals seen in Beijing normally yellowish brown overall, with broad wings, rounded head, short tail, and dark "carpal patches" under the wings. Subadult has distinct longitudinal stripes on the belly.

Often inhabits mountain forests, foothills, plains, and grasslands; frequents open fields, farmlands, wastelands, villages, and similar habitats in winter. Usually solitary or in pairs but gathers in flocks during migration. Feeds mostly on rodents, frogs, also small and medium-sized birds.

Occurs in most regions in eastern China. A passage migrant and winter visitor in Beijing where it is common in suburban areas such as Yanqing, Miyun, and Huairou.

051

中文名 **红隼**

学　名 *Falco tinnunculus*

英文名 Common Kestrel

隼形目
FALCONIFORMES

隼科
Falconidae

红隼是一种中型猛禽。雄鸟头灰色，脸颊白色；背部砖红色，具黑色点斑；腹部黄褐色，具黑色斑纹；尾灰色，尾羽次端黑色。雌鸟头部、背部红褐色。亚成鸟似雌鸟，未成年雄鸟头灰色较明显。

红隼常栖息于农田、村落、山地森林、草原、旷野等地带，部分个体在城市中繁殖。常单独活动。可在空中悬停观察地面情况，如发现猎物，快速俯冲捕捉。主要以鼠类、小型鸟类为食。

红隼分布于我国多数地区。北京城区及郊区均可见，为留鸟、旅鸟及冬候鸟。

	北京市重点保护 Beijing priority conservation	国家重点保护 State priority conservation	IUCN
保护级别 Protected-level	——	二级（II）	无危（LC）2016.10.1

A medium-sized raptor. Male has gray head, white cheek, brick-red back with black spots, yellow-brown belly with black markings, gray tail with broad black subterminal band. Female has reddish-brown head and back. Subadult similar to female, while immature male has more pronounced gray head.

Often inhabits farmlands, villages, mountain forests, grasslands, and open fields. Some individuals breed in urban areas. Often solitary. Hovers mid-air while observing the ground; quickly dives down to capture prey. Feeds mostly on rodents and small birds.

Occurs in most regions in China. A resident, passage migrant, and winter visitor in Beijing where it can be seen in urban and suburban areas.

中文名 红脚隼

学　名 *Falco amurensis*

英文名 Eastern Red-footed Falcon

隼形目
FALCONIFORMES

隼科
Falconidae

红脚隼是一种小型猛禽。雌雄差异大。雄鸟头、胸、背深灰色，尾下覆羽红色明显。雌鸟头灰色，脸颊白色，翼下密布斑纹，尾下覆羽红色较明显。亚成鸟似雌鸟，眼圈和虹膜橙黄色，上体具浅褐色羽缘。

红脚隼常栖息于平原、低山、丘陵、农田等环境。非繁殖季成群活动。飞行速度较快，可在空中追捕昆虫，亦可在空中悬停寻觅食物。主要以蝗虫、蚱蜢、螽斯、金龟子等昆虫为食。

红脚隼分布于我国东部多数地区。北京城区及郊区均可见，为旅鸟和夏候鸟。

北京市重点保护 Beijing priority conservation	国家重点保护 State priority conservation	IUCN
——	二级（II）	无危（LC）2016.10.1

A small raptor, sexes different. Male dark grey in head, chest, and back; undertail-coverts distinct red. Female has grey head, white check, dense stripes under wings, and reddish undertail-coverts. Juvenile similar to female, eye-ring and iris orange; feather edge light brown on the upperparts.

Often inhabits environments like plains, low mountains, hills, and farmlands. Form flocks in the non-breeding season. Flies rapidly. Feeds mostly on insects such as locusts, grasshoppers, katydids, and chafers. Can chase insects in the air or hover mid-air while searching for food.

Occurs in most regions in eastern China. A passage migrant and summer visitor in Beijing where it can be seen in urban and suburban areas.

053

中文名 **燕隼**
学　名 *Falco subbuteo*
英文名 Hobby

隼形目
FALCONIFORMES
隼科
Falconidae

燕隼是一种中型猛禽。雌雄相似。双翼窄长，成鸟黑色“头盔”明显，胸部点状斑纹明显，尾下覆羽红色明显。亚成鸟与成鸟相似，尾下覆羽皮黄色。

燕隼常栖息于开阔平原、稀疏林地、旷野、农田等环境。常单独或成对活动。飞行速度快而敏捷，可在空中追击家燕、蜻蜓等。主要以小型鸟类、昆虫为食。

燕隼分布于我国多数地区。北京延庆、密云、怀柔等地可见，为旅鸟和夏候鸟。

北京市重点保护 Beijing priority conservation	国家重点保护 State priority conservation	IUCN
——	二级（II）	无危（LC）2016.10.1

亚成鸟 Immature

A medium-sized raptor, sexes similar, with slender and elongated wings. Adult has black "helmet" on the head, and noticeable spotted chest; undertail-coverts distinct red. Subadult similar to adult but undertail-coverts light yellow.

Inhabits plains, sparse woodlands, wilderness areas, and farmlands. Usually solitary or in pairs. Fast and agile; capable of chasing swallows, dragonflies, and other prey in the air. Feeds mostly on small birds and insects.

Occurs in most regions in China. A passage migrant and summer visitor in Beijing where it can be seen in places such as Yanqing, Miyun, and Huairou.

中文名 **黑卷尾**
学　名 *Dicrurus macrocercus*
英文名 Black Drongo

雀形目
PASSERIFORMES
卷尾科
Dicruridae

黑卷尾是一种中型鸣禽。雌雄相似；成鸟整体黑色，似小型乌鸦，虹膜红褐色，喙黑色，喙基有刚毛；尾羽较长，分叉深，尾端略上卷。

黑卷尾常栖息于开阔平原、旷野、乡村等地带。常单独或成对活动。繁殖期性凶猛，常驱逐进入其领地的其他鸟类，飞行敏捷，擅在空中捕食昆虫。

黑卷尾分布于我国多数地区。北京延庆、怀柔等地可见，为旅鸟和夏候鸟。

北京市重点保护 Beijing priority conservation	国家重点保护 State priority conservation	IUCN
是（YES）	—	无危（LC）2016.10.1

幼鸟 Juvenile

A medium-sized songbird, sexes similar. Adult blackish overall, like a small crow; iris reddish-brown; beak black with the base bristled; long tail deeply forked with the end slightly curved upwards.

Inhabits plains, wilderness, countryside. Usually solitary or in pairs. Pugnacious during breeding, frequently chases away other birds in territory. Swift and agile; sallies to catch prey on the wing.

Occurs in most regions in China. A passage migrant and summer visitor in Beijing where it can be seen in places such as Yanqing and Huairou.

055 中文名 红尾伯劳

学　名 *Lanius cristatus*

英文名 Brown Shrike

雀形目
PASSERIFORMES

伯劳科
Laniidae

红尾伯劳是一种中型鸣禽。雌雄相似。成鸟头顶红褐色或灰色，头部具一黑色贯眼纹，喙黑色带钩，上体红褐色，下体黄褐色，尾羽红棕色。雌鸟似雄鸟，胸腹部带横纹。幼鸟似雌鸟，但胸腹部多鳞状斑纹。

红尾伯劳常栖息于开阔平原、稀疏林地、农田、果园等地带。常单独活动。性凶猛，喜在开阔地带的高处寻找食物。主要以昆虫、小型蛙类、鼠类等为食。

红尾伯劳分布于我国东部多数地区。北京延庆、怀柔及城区公园可见，为旅鸟和夏候鸟。

北京市重点保护 Beijing priority conservation	国家重点保护 State priority conservation	IUCN
是（YES）	—	无危（LC）2016.10.1

A medium-sized songbird. Sexes similar with female marked by transverse patterns on the chest and belly. Adult reddish-brown or gray in crown with black eye-stripe; black beak hooked; upperparts reddish-brown; underparts yellowish-brown; rectrices reddish brown. Juvenile similar to female, but with scaly markings on the chest and belly.

Inhabits plains, sparse woodlands, farmland, orchards. Solitary and fierce. Prefers to perch high in open land to hunt. Feeds mostly on insects, small frogs, rodents, etc.

Occurs in most regions in eastern China. A passage migrant and summer visitor in Beijing where it can be seen in Yanqing, Huairou, and urban parks.

056

中文名 **楔尾伯劳**
学　名 *Lanius sphenocercus*
英文名 Chinese Gray Shrike

雀形目
PASSERIFORMES
伯劳科
Laniidae

楔尾伯劳是一种中型鸣禽。雌雄相似，个体相对较大。成鸟头顶灰色，眉纹白色，贯眼纹黑色；背部灰色，具黑色、白色翼斑；胸腹部偏白色，腰部灰色；尾羽黑色，尾羽两侧白色。

楔尾伯劳常栖息于开阔平原、农田、果园、低山丘陵地带。常单独活动。性凶猛，喜在开阔地带的高处寻找食物，可在空中悬停。主要以昆虫、鼠类等为食。

楔尾伯劳分布于我国东部多数地区。北京延庆、密云等地可见，为旅鸟和冬候鸟。

北京市重点保护 Beijing priority conservation	国家重点保护 State priority conservation	IUCN
是（YES）	——	无危（LC）2016.10.1

成鸟 Adult

A medium-sized songbird, large in shrikes, sexes similar. Adult has gray crown, white supercilium, black eye-stripe; gray back with black-and-white bars on the wings; whitish chest and belly; gray rump; black rectrices with outer tail feathers white.

Commonly inhabits plains, farmlands, orchards, and low hills. Often solitary. Fierce by nature; prefers to hunt from elevated positions in open areas; able to hover. Feeds mostly on insects, rodents, etc.

Occurs in most regions in eastern China. A passage migrant and winter visitor in Beijing that can be seen in places such as Yanqing and Miyun.

中文名 **灰喜鹊**
学　名 *Cyanopica cyanus*
英文名 Azure-winged Magpie

雀形目
PASSERIFORMES
鸦科
Corvidae

灰喜鹊是一种大型鸣禽。雌雄相似。头部黑色，喙黑色，喉部、胸腹部、背部灰色，两翼蓝色；尾较长，呈蓝色，尾端白色。

灰喜鹊常栖息于城区及低山地带的次生林、人工林及城市公园。常成群活动。主要以昆虫、植物果实等为食。

灰喜鹊分布于我国东部多数地区。北京城区及郊区常见，为留鸟。

北京市重点保护 Beijing priority conservation	国家重点保护 State priority conservation	IUCN
——	——	无危（LC）2016.10.1

成鸟 Adult

A large songbird, sexes similar with black head and beak; gray throat, back, chest and belly; azure wings; long powder-blue tail and white terminal band.

Often inhabits urban and lower mountain territories such as parks, secondary forests and plantation forests. Usually in flocks. Feeds mostly on insects and fruits.

Occurs in most regions in eastern China. A resident in Beijing where it is common in urban and suburban areas.

中文名 **红嘴蓝鹊**
学　名 *Urocissa erythroryncha*
英文名 Red-billed Blue Magpie

雀形目
PASSERIFORMES

鸦科
Corvidae

成鸟 Adult

红嘴蓝鹊是一种大型鸣禽。雌雄相似。头部黑色，头顶白色延伸至枕部，喙红色，喉部、胸部黑色，腹部白色，背部蓝色，跗跖红色；尾羽甚长，呈蓝色，尾两侧具黑色、白色斑纹，尾端白色。

红嘴蓝鹊常栖息于低山及平原地带的阔叶林、混交林等环境。常成对或成小群活动。性凶猛，有时会捕捉其他鸟类或小型兽类为食。

红嘴蓝鹊分布于我国多数地区。北京郊区及城区常见，为留鸟。

北京市重点保护 Beijing priority conservation	国家重点保护 State priority conservation	IUCN
是（YES）	——	无危（LC）2018.8.7

A large songbird, sexes similar. Head black with white color extending from the cap to the nape; beak red; throat and chest black; belly white; back blue; tarsi red; long blue tail marked with black-and-white bars on the outer tail feathers and white terminal band.

Often inhabits broadleaf and mixed forests in low mountains and plains. Usually in pairs or small flocks. Aggressive, sometimes preys on other birds and small mammals.

Occurs in most regions in China. A resident in Beijing where it is common in urban and suburban areas.

中文名 **喜鹊**
学　名 *Pica serica*
英文名 Oriental Magpie

雀形目
PASSERIFORMES

鸦科
Corvidae

喜鹊是一种大型鸣禽。雌雄相似。头部黑色，喙黑色，喉部、胸黑色，腹部白色，背部黑色，两翼具金属蓝绿色光泽及一道白斑，腰部灰白色；尾较长，具金属蓝绿色光泽。

喜鹊适应能力强，可在各海拔的林地、农田、湿地等环境活动，尤好栖息于有人居住的环境。单独或成群活动，冬季有时成数十只的大群。主要以昆虫、植物果实、种子等为食。

喜鹊分布于我国除西藏和新疆外的多数地区。北京城区及郊区常见，为留鸟。

保护级别 Protected-level	北京市重点保护 Beijing priority conservation	国家重点保护 State priority conservation	IUCN
	—	—	无危（LC）2016.10.1

成鸟 Adult

A large songbird, sexes similar. Head and bill black, as well as the throat, chest, and back; belly white; wings have metallic blue-green luster and a white band; rump grayish white; tail comparatively long with metallic blue-green luster.

Adaptable to various habitats at different elevations including forests, farmlands, wetlands; prefers environments with human settlements. Seen solitary or in flocks, which sometimes contain dozens of individuals in winter. Feeds mostly on insects, fruits, and seeds.

Occurs in most regions in China except for Xizang and Xinjiang. A resident in Beijing where it is common in urban and suburban areas.

060 中文名 达乌里寒鸦

学　名 *Corvus dauuricus*

英文名 Daurian Jackdaw

雀形目
PASSERIFORMES

鸦科
Corvidae

成鸟 Adult

达乌里寒鸦是一种大型鸣禽。雌雄相似。成鸟头部黑色，喙黑色、较短，颈部白色向胸侧延伸，喉部、胸黑色，腹部白色，背部及两翼、尾黑色。亚成鸟整体黑色。

达乌里寒鸦常栖息于开阔平原、稀疏林地、农田等环境。冬季常成大群活动。主要以植物果实、种子及垃圾堆中的残渣为食。

达乌里寒鸦分布于我国除海南以外的多数地区。北京郊区可见，城区偶见，为冬候鸟和旅鸟。

北京市重点保护 Beijing priority conservation	国家重点保护 State priority conservation	IUCN
—	—	无危（LC）2016.10.1

A large songbird, sexes similar. Adult has black head with relatively short black bill; white on the neck extending to the side of the chest; throat and chest black; belly white; back, wings and tail black. Subadult black overall.

Inhabits plains, sparse woodlands, farmland and cultivated environments. Often form large flocks in winter. Feeds mostly on fruits, seeds, and debris in garbage piles.

Occurs in most regions of China except for Hainan. A winter visitor and passage migrant in Beijing where it can be seen in suburban and occasionally in urban areas.

061

中文名 **小嘴乌鸦**
学　名 *Corvus corone*
英文名 Carrion Crow

雀形目
PASSERIFORMES
鸦科
Corvidae

成鸟 Adult

小嘴乌鸦是一种大型鸣禽。雌雄相似，个体较大的乌鸦。整体黑色，额头较平，喙较直。

小嘴乌鸦常栖息于开阔平原地带的稀疏林地、农田、城市等环境。冬季常成大群活动，常夜宿在北京城区，白天至郊区觅食。主要以垃圾堆中的残渣、农耕地中的粮食等为食。

小嘴乌鸦分布于我国多数地区。北京城区、郊区常见，为留鸟、冬候鸟和旅鸟。

北京市重点保护 Beijing priority conservation	国家重点保护 State priority conservation	IUCN
——	——	无危（LC）2016.10.1

成鸟 Adult

A large songbird, sexes similar. Relatively large in crows; black overall with flatter forehead and straighter bill.

Often inhabits sparse woodlands, farmlands, and urban areas in plains. Usually in large flocks in winter. In Beijing, often roosts in the urban areas at night and ventures to the suburban areas during the day for food. Feeds mostly on leftover food from garbage dumps and grains from cultivated fields.

Occurs in most regions of China. A resident, winter visitor, and passage migrant in Beijing where it is common in urban and suburban areas.

062 中文名 大嘴乌鸦

学　名 *Corvus macrorhynchos*

英文名 Large-billed Crow

雀形目 PASSERIFORMES

鸦科 Corvidae

成鸟 Adult

大嘴乌鸦是一种大型鸣禽。雌雄相似，个体较大的乌鸦。整体黑色，额头明显隆起，喙较粗壮。

大嘴乌鸦常栖息于平原、丘陵、山区林地等环境。冬季常成群活动。主要以昆虫、植物果实、种子、农作物等为食。

大嘴乌鸦分布于我国除西北地区以外的多数地区。北京城区、郊区常见，为留鸟。

北京市重点保护 Beijing priority conservation	国家重点保护 State priority conservation	IUCN
—	—	无危（LC）2016.10.1

成鸟 Adult

A large songbird, sexes similar. Relatively large in crows; black overall with forehead prominently bulging; beak thick and stout.

Commonly inhabits environments like plains, hills, and mountainous forests. Usually in flocks in winter. Feeds mostly on insects, fruits, seeds, and agricultural crops.

Occurs in most regions of China except for the northwest. A resident in Beijing where it is common in urban and suburban areas.

中文名 黄腹山雀

学　名 *Pardaliparus venustulus*

英文名 Yellow-bellied Tit

雀形目
PASSERIFORMES

山雀科
Paridae

黄腹山雀是一种小型鸣禽。雌雄差异大。雄鸟头部黑色，脸颊、枕部白色，喉部黑色；上体黑灰色，具两条白色翼斑；下体黄色；黑色尾羽较短，尾端白色。雌鸟整体色浅，头部灰色，脸颊灰白色，喉部黄绿色。

黄腹山雀常栖息于山区或平原地带的针叶林、混交林等地带。非繁殖季节常成群活动。主要以昆虫、植物果实、种子为食。

黄腹山雀分布于我国东部多数地区。北京城区及郊区均可见，为留鸟和旅鸟。

北京市重点保护 Beijing priority conservation	国家重点保护 State priority conservation	IUCN
是（YES）	——	无危（LC）2016.10.1

A small songbird, sexes different. Male has black head with cheek and nape white; throat black; Upperparts blackish-gray with two white wing bars; underparts yellow; black rectrices relatively short with terminal band white. Female duller overall; head gray with cheek grayish-white; throat yellowish-green.

Commonly inhabits mountainous and plain areas with coniferous forests, mixed forests. Usually in flocks during the non-breeding season. Feeds mostly on insects, fruits, and seeds.

Occurs in most regions in eastern China. A resident and passage migrant in Beijing where it can be seen in urban and suburban areas.

中文名 **沼泽山雀**
学　名 *Poecile palustris*
英文名 Marsh Tit

雀形目
PASSERIFORMES

山雀科
Paridae

沼泽山雀是一种小型鸣禽。雌雄相似。成鸟头顶、喉部黑色，脸颊白色，上体灰褐色，飞羽羽缘灰白色，下体灰褐色，尾羽灰褐色。

沼泽山雀常栖息于低山或平原地带的针叶林、混交林、城区公园、果园等环境。非繁殖季节常成小群在树冠层活动。主要以昆虫、植物果实、种子为食。

沼泽山雀分布于我国东部多数地区。北京城区及近郊可见，为留鸟。

北京市重点保护 Beijing priority conservation	国家重点保护 State priority conservation	IUCN
——	——	无危（LC）2016.10.1

A small songbird, sexes similar. Adult has black cap and throat; white cheek; upperparts grayish-brown with the edge of the flight feathers grayish-white; underparts and tail grayish-white.

Often inhabits low mountains and plains with coniferous forests, mixed forests, urban parks, and orchards. Usually in small flocks and forages in the canopy during the non-breeding season. Feeds mostly on insects, fruits, and seeds.

Occurs in most regions in eastern China. A resident in Beijing where it can be seen in urban and suburban areas.

065 中文名 大山雀

学　名 *Parus minor*

英文名 Japanese Tit

雀形目
PASSERIFORMES

山雀科
Paridae

大山雀是一种小型鸣禽。雌雄相似。成鸟头顶黑色，脸颊白色，喉部黑色并向下延伸至腹部形成一道黑色纵纹，胸腹部白色，背部灰色或灰绿色，翼上具一道白色斑纹，尾羽灰黑色。雄鸟腹部黑色纵纹较雌鸟宽。

大山雀常栖息于山区或平原地带的针叶林、混交林、城市公园、果园等地带。常单独或成小群活动。主要以昆虫、植物果实、种子为食。

大山雀分布于我国除西北地区以外的多数地区。北京城区及延庆、密云、怀柔可见，为留鸟。

北京市重点保护 Beijing priority conservation	国家重点保护 State priority conservation	IUCN
—	—	无危（LC）2016.10.1

成鸟 Adult

A small songbird, sexes similar. Adult has black crown and white cheek; throat black with the color running down to the belly, creating a black longitudinal line; chest and belly white; back gray or grayish-green; wings marked with a white bar; tail feathers grayish-black. The black longitudinal line is wider in male than in female.

Inhabits mountainous and plain areas with coniferous forests, mixed forests, urban parks, and orchards. Usually solitary or in small flocks. Feeds mostly on insects, fruits, and seeds.

Occurs in most regions of China except for the northwest. A resident in Beijing where it can be seen in urban areas, Yanqing, Miyun, and Huairou.

中文名 **云雀**
学　名 *Alauda arvensis*
英文名 Eurasian Skylark

雀形目
PASSERIFORMES

百灵科
Alaudidae

云雀是一种中型鸣禽。雌雄相似。成鸟整体棕黄色，眼先和眉纹棕白色，耳羽棕黄色，头顶具一短羽冠，喙黄褐色；上体棕色，具黑色斑纹；胸侧棕色，具黑色点斑；腹部米黄色；尾棕黄色。

云雀常栖息于开阔的平原、草地、农田等地带。繁殖期会从地面直飞至空中炫耀飞行，非繁殖期常成小群活动。主要以昆虫、植物果实、种子为食。

云雀分布于我国多数地区。北京延庆、密云可见，为旅鸟和冬候鸟。

北京市重点保护 Beijing priority conservation	国家重点保护 State priority conservation	IUCN
——	二级（II）	无危（LC）2018.8.9

A medium-sized songbird, sexes similar. Adult brownish-yellow overall, with brownish-white lore and supercilium; ear-coverts brownish-yellow; crest short; bill yellowish-brown; upperparts brown with black markings; chest side brown with black spots; belly beige; tail brownish-yellow.

Inhabits environments such as plains, grasslands, farmlands, and cultivated areas. Flies straight up from the ground into the air to display during the breeding season. Often in flocks when not breed. Feeds mostly on insects, fruits, and seeds.

Occurs in most regions of China. A passage migrant and winter visitor in Beijing where it can be seen in Yanqing and Miyun.

中文名 东方大苇莺

学　名 *Acrocephalus orientalis*

英文名 Oriental Reed Warbler

雀形目
PASSERIFORMES

苇莺科
Acrocephalidae

成鸟 Adult

东方大苇莺是一种中型鸣禽。雌雄相似。成鸟整体黄褐色，眉纹白色，耳羽黄褐色，上喙深褐色，下喙粉褐色，喉部偏白色，上体黄褐色，下体米黄色，尾黄褐色。

东方大苇莺常栖息于平原地带的湖泊、河流、水塘等水域中茂密的芦苇丛环境。繁殖期常单独或成对活动，性活泼，清晨常立于芦苇或树枝高处鸣叫。主要以昆虫为食。

东方大苇莺分布于我国东部多数地区。北京延庆、密云、怀柔及城区可见，为旅鸟和夏候鸟。

北京市重点保护 Beijing priority conservation	国家重点保护 State priority conservation	IUCN
是（YES）	——	无危（LC）2016.10.1

A medium-sized songbird, sexes similar. Adult yellowish-brown overall with supercilium white; ear-coverts yellowish-brown; upper mandible dark brown, lower mandible pale brown; throat whitish; upperparts yellowish-brown, underparts beige; tail yellowish-brown.

Inhabits dense reed beds of lakes, rivers, and ponds in plains. Usually solitary or in pairs during the breeding season. Active, often sings from high perches in the early morning. Feeds mostly on insects.

Occurs in most regions in eastern China. A passage migrant and summer visitor in Beijing where it can be seen in Yanqing, Miyun, Huairou, and urban areas.

068

中文名 **家燕**
学　名 *Hirundo rustica*
英文名 Barn Swallow

雀形目
PASSERIFORMES
燕科
Hirundinidae

成鸟 Adult

幼鸟 Juvenile

家燕是一种中型鸣禽。雌雄相似。成鸟头顶、颈部、背部带蓝黑色金属光泽，喙黑色，翼近黑色，前额、喉部红褐色，胸、腹部白色，近黑色尾呈叉状。繁殖期雄鸟尾较雌鸟长。

家燕常栖息于城区、农田、荒野、湿地等环境。繁殖期常单独或成对活动。在空中飞行时快速而敏捷。主要以昆虫为食。

家燕分布于我国多数地区。北京城区及郊区常见，为夏候鸟和旅鸟。

北京市重点保护 Beijing priority conservation	国家重点保护 State priority conservation	IUCN
是（YES）	——	无危（LC）2018.11.15

成鸟 Adult

A medium-sized songbird, sexes similar. Adult has bluish-black metallic luster on the crown, neck, and back; beak black; wings near black; forehead and throat reddish-brown; chest and belly white; blackish tail forked. Male has longer tail during the breeding season.

Often inhabits environments like urban areas, farmlands, wastelands, wetlands. Usually solitary or in pairs during the breeding season. Swift and agile. Feeds mostly on insects.

Occurs in most regions in China. A summer visitor and passage migrant in Beijing where it is common in urban and suburban areas.

中文名 **金腰燕**

学　名 *Cecropis daurica*

英文名 Red-rumped Swallow

雀形目 PASSERIFORMES

燕科 Hirundinidae

成鸟 Adult

金腰燕是一种中型鸣禽。雌雄相似。成鸟头顶、背部带蓝色金属光泽，喙黑色，两翼及尾羽近黑色，颈侧橘红色；喉部、胸、腹部白色，具黑色纵纹；近黑色尾呈叉状。

金腰燕常栖息于城区、农田、荒野等环境。繁殖期常单独或成对活动，有时与家燕混群活动。在空中飞行敏捷。主要以昆虫为食。

金腰燕分布于我国除西北以外的多数地区。北京城区及郊区常见，为夏候鸟和旅鸟。

北京市重点保护 Beijing priority conservation	国家重点保护 State priority conservation	IUCN
是（YES）	——	无危（LC）2016.10.1

A medium-sized songbird, sexes similar. Adult has blue metallic luster on the crown and back; beak black; wings and tail near black; neck side orange-red; throat white, chest, and belly marked with black streaks; blackish tail forked.

Often inhabits environments like urban areas, farmland, and wilderness. Usually solitary or in pairs during the breeding season. Sometimes form mixed-species flocks with Barn Swallows. Swift and agile. Feeds mostly on insects.

Occurs in most regions in China except the northwest. A summer visitor and passage migrant in Beijing where it is common in urban and suburban areas.

中文名 **白头鹎**
学　名 *Pycnonotus sinensis*
英文名 Light-vented Bulbul

雀形目
PASSERIFORMES
鹎科
Pycnonotidae

白头鹎是一种中型鸣禽。雌雄相似。成鸟头顶黑色，眼后白斑延伸至脑后，耳羽灰白色，喉部、胸腹部灰白色，胸侧沾灰色，背部、两翼及尾羽灰绿色。

白头鹎常栖息于开阔平原、城区公园、农田、果园、低山丘陵地带。性胆大，不甚惧人。非繁殖期常成群活动。主要以昆虫、植物果实为食。

白头鹎分布于我国除西北以外的多数地区。北京城区及郊区常见，为留鸟、旅鸟和冬候鸟。

保护级别 Protected-level	北京市重点保护 Beijing priority conservation	国家重点保护 State priority conservation	IUCN
	——	——	无危（LC）2018.8.9

幼鸟 / 左，成鸟 / 右 Juvenile/left, Adult/right

A medium-sized songbird, sexes similar. Adult has black crown; white patches behind the eyes extending to the back of the head; ear-coverts grayish-white, as well as the throat, chest and belly; chest side grayish; back, wings and tail grayish-green.

Often inhabits plains, urban parks, farmland, orchards and low hills. Bold and unafraid of people. Usually in flocks during the non-breeding season. Feeds mostly on insects and fruits.

Occurs in most regions in China except for the northwest. A resident, passage migrant and winter visitor in Beijing where it is common in the urban and suburban areas.

中文名 **黄眉柳莺**
学　名 *Phylloscopus inornatus*
英文名 Yellow-browed Warbler

雀形目
PASSERIFORMES
柳莺科
Phylloscopidae

成鸟 Adult

黄眉柳莺是一种小型鸣禽。雌雄相似。成鸟整体黄绿色，头部具一米黄色眉纹，喙深褐色，下喙基肉色，绿色翼上具两道米黄色翼斑，胸腹部灰白色；尾羽较短，绿色。

黄眉柳莺常栖息于开阔平原、城区公园、中低海拔的针叶林、混交林、灌丛等环境。性活泼，常活动于树冠层。主要以昆虫、植物果实为食。

黄眉柳莺分布于我国多数地区。北京城区及郊区常见，为旅鸟。

北京市重点保护 Beijing priority conservation	国家重点保护 State priority conservation	IUCN
——	——	无危（LC）2018.11.21

A small songbird, sexes similar. Adult yellowish-green overall with yellowish-white supercilium; bill dark brown with the lower base flesh-colored; wings green marked with two flesh-colored bars; chest and belly grayish-white; short tail green.

Often inhabits environments such as plains, urban parks, as well as coniferous forests, mixed forests, and shrublands at middle and low elevations. Active, often seen in canopy. Feeds mostly on insects and fruits.

Occurs in most regions of China. A passage migrant in Beijing where it is common in both urban and suburban areas.

中文名 **黄腰柳莺**

学　名 *Phylloscopus proregulus*

英文名 Pallas's Leaf Warbler

雀形目
PASSERIFORMES

柳莺科
Phylloscopidae

成鸟 Adult

黄腰柳莺是一种小型鸣禽。雌雄相似。成鸟头部具一黄色眉纹，眉纹前端色重；头顶具一清晰的黄色顶冠纹；绿色翼上具两道黄色翼斑，胸腹部灰白色，腰部黄色；尾羽较短，绿色。

黄腰柳莺常栖息于开阔平原、城区公园、中低海拔的山区林地等环境。性活泼，常活动于树冠层，有时与其他柳莺或山雀混群活动。主要以昆虫、植物果实为食。

黄腰柳莺分布于我国除西北以外的多数地区。北京城区及郊区常见，为旅鸟和冬候鸟。

北京市重点保护 Beijing priority conservation	国家重点保护 State priority conservation	IUCN
是（YES）	——	无危（LC）2016.10.1

成鸟 Adult

A small songbird, sexes similar. Adult has yellow supercilium with the color more intense at the beginning; clear yellow crown-stripe; green wings marked with two yellow bars; chest and belly grayish-white; rump yellow; green tail relatively short.

Often inhabits plains, urban parks, also mountainous forests at middle and low elevations. Active, often seen in canopy. Sometimes joins mix-species flocks with other warblers and tits. Feeds mostly on insects and fruits.

Occurs in most regions of China except for the northwest. A passage migrant and winter visitor in Beijing where it is common in urban and suburban areas.

中文名 **银喉长尾山雀**
学　名 *Aegithalos glaucogularis*
英文名 Silver-throated Bushtit

雀形目
PASSERIFORMES

长尾山雀科
Aegithalidae

银喉长尾山雀是一种体小而圆的鸣禽。雌雄相似。成鸟头顶白色，头顶两侧黑色，脸颊灰白色，喙黑色；喉部白色，中央黑色；胸腹部沾粉色，背部灰黑色，两翼黑色，尾灰黑色。

银喉长尾山雀常栖息于开阔平原、中低海拔的针叶林、混交林、灌丛等环境。有时与其他山雀混群。主要以昆虫、植物果实为食。

银喉长尾山雀分布于我国东部多数地区。北京城区及郊区常见，为留鸟。

北京市重点保护 Beijing priority conservation	国家重点保护 State priority conservation	IUCN
是（YES）	——	无危（LC）2016.10.1

成鸟 Adult

A small and round songbird, sexes similar. Adult has white crown with black lateral crown-stripe; cheek grayish-white; bill black; throat white with a dark patch in the middle; chest and belly pinkish; back grayish-black; wings and tail black.

Inhabits plains, also coniferous forests, mixed forests, and shrubs at middle and low elevations. Sometimes joins flocks with other tits. Feeds mostly on insects and fruits.

Occurs in most regions of eastern China. A resident in Beijing where it is common in urban and suburban areas.

中文名 **棕头鸦雀**
学　名 *Sinosuthora webbiana*
英文名 Vinous-throated Parrotbill

雀形目
PASSERIFORMES
莺鹛科
Sylviidae

成鸟 Adult

棕头鸦雀是一种体小尾长的鸣禽。雌雄相似。成鸟头部棕色，虹膜黑色；喙厚实，呈深灰色，喙尖米黄色；背部灰褐色，两翼棕红色，胸腹部灰褐色；尾羽较长，灰褐色。

棕头鸦雀常栖息于平原地区和中低海拔山区的次生林、灌丛、竹林、芦苇丛等环境。性大胆，不甚惧人，常集群活动。主要以昆虫为食。

棕头鸦雀分布于我国东部多数地区。北京城区及郊区常见，为留鸟。

保护级别 Protected-level	北京市重点保护 Beijing priority conservation	国家重点保护 State priority conservation	IUCN
	是（YES）	—	无危（LC）2018.8.9

成鸟 Adult

A small songbird with long tail, sexes similar. Adult has brown head and black iris; thick bill dark gray with beige tip; back, chest and belly grayish-brown; wings reddish-brown; tail grayish-brown and rather long.

Often inhabits plains, also secondary forests, shrubs, bamboo groves, and reed beds at low and middle altitude mountains. Bold and unafraid of people. Often in flocks. Feeds mostly on insects.

Occurs in most regions in eastern China. A resident in Beijing where it is common in urban and suburban areas.

中文名 **山噪鹛**
学　名 *Pterorhinus davidi*
英文名 Plain Laughingthrush

雀形目
PASSERIFORMES
噪鹛科
Leiothrichidae

成鸟 Adult

山噪鹛是一种中型鸣禽。雌雄相似。整体棕色，虹膜红褐色，黄色喙较长而下弯，两翼较短，翼缘灰褐色；棕色尾羽较长，端部灰褐色。

山噪鹛常栖息于山区及近山平原地区的灌丛、低矮树林等环境。性大胆而活跃，不甚惧人，非繁殖季常集小群活动。喜在地面觅食，主要以昆虫为食。

山噪鹛分布于我国华北、华南、西南等地区。北京郊区较常见，为留鸟。

北京市重点保护 Beijing priority conservation	国家重点保护 State priority conservation	IUCN
是（YES）	——	无危（LC）2016.10.1

成鸟 Adult

A medium-sized songbird, sexes similar. Grayish brown overall; chestnut iris; yellow bill rather long and down-curved; rather short wings with gray wing panel; tail feathers brown and rather long, with grayish brown tips.

Inhabits environments such as shrublands and low-stature forests in plains and mountainous areas. Bold, active, and unafraid of people. Form small flocks in the non-breeding season. Likes foraging on the ground; feeds mostly on insects.

Occurs in regions such as north, south, and southwestern China. A resident in Beijing where it is rather common in suburban areas.

中文名 **黑头䴓**
学　名 *Sitta villosa*
英文名 Chinese Nuthatch

雀形目
PASSERIFORMES
䴓科
Sittidae

黑头䴓是一种小型鸣禽。雌雄相似。头顶及贯眼纹黑色，耳羽、下颏、眉纹白色，胸腹部棕黄色，背部、腰部蓝灰色，两翼近黑色，蓝灰色尾羽甚短。雌鸟似雄鸟，但头顶及贯眼纹色浅，为灰色。

黑头䴓常栖息于山区及近山平原地区的针叶林、针阔混交林等环境。常单独或成对活动，擅在树干螺旋式上下攀爬。主要以昆虫为食。

黑头䴓分布于我国华北、华南、西南等地区。北京山区及国家植物园较常见，为留鸟。

北京市重点保护 Beijing priority conservation	国家重点保护 State priority conservation	IUCN
是（YES）	—	无危（LC）2016.10.1

成鸟 Adult

A small songbird, sexes similar. Black crown and eye-stripe; white auriculars, chin and supercilium; buff chest and belly; bluish gray back and rump; wings nearly black; tail bluish gray and rather short; female similar to male, but with duller crown and eye-stripe.

Often inhabits environments such as coniferous forests and mixed forests in plains and mountainous areas. Usually solitary or in pairs. Adept at spiraling up and down along tree trunks. Feeds mostly on insects.

Occurs in regions such as north, south, and southwestern China. A resident in Beijing where it is common in the China National Botanical Garden and mountainous areas.

中文名 **鹪鹩**

学　名 *Troglodytes troglodytes*

英文名 Eurasian Wren

雀形目
PASSERIFORMES

鹪鹩科
Troglodytidae

成鸟 Adult

鹪鹩是一种小型鸣禽。雌雄相似。整体棕褐色，密布黑褐色横纹；头部具一黄褐色眉纹；喙较细长，上喙黑色，下喙黄褐色。

鹪鹩常栖息于山区及平原地区的林地、灌丛及多石溪流环境。性隐秘，喜在阴暗处跳动，常单独活动。主要以昆虫为食。

鹪鹩分布于我国多数地区。北京于国家植物园、奥林匹克森林公园及山区较常见，为垂直迁徙的留鸟。

北京市重点保护 Beijing priority conservation	国家重点保护 State priority conservation	IUCN
—	—	无危（LC）2018.8.9

成鸟 Adult

A small songbird, sexes similar. Brown overall with densely packed dark barring; pale buff supercilium; slender bill with black maxilla and yellowish brown lower mandible.

Often inhabits woodlands, shrublands, and rocky streams in plains and mountainous areas. Secretive, likes hopping around in dark places; usually solitary. Feeds mostly on insects.

Occurs in most regions in China. An elevational migrant in Beijing where it is rather common in the China National Botanical Garden, the Olympic Forest Park, and mountainous areas.

中文名 **灰椋鸟**

学　名 *Spodiopsar cineraceus*

英文名 White-cheeked Starling

雀形目
PASSERIFORMES

椋鸟科
Sturnidae

灰椋鸟是一种中型鸣禽。雌雄相似。头部黑色，耳羽白色，黄色喙较细长，背部、两翼、胸部灰色，腹部偏白色，腰部白色，灰色尾羽较短。飞行时两翼似三角形。

灰椋鸟常栖息于平原地区的公园绿地、农田等环境。非繁殖季常成群活动，有时与丝光椋鸟、黑尾蜡嘴雀等混群活动。主要以昆虫为食。

灰椋鸟分布于我国除西北以外的多数地区。北京城区各公园较常见，为夏候鸟、旅鸟及冬候鸟。

北京市重点保护 Beijing priority conservation	国家重点保护 State priority conservation	IUCN
——	——	无危（LC）2016.10.1

成鸟 Adult

A medium-sized songbird, sexes similar. Black head with white auriculars; yellow and slender bill; gray back, wings, and chest; white belly and rump; gray, rather short tail; The wings are triangular in flight.

Often inhabits environments such as green spaces in parks and agricultural fields in plains; form flocks in the non-breeding season, sometimes with Red-billed Starlings and Chinese Grosbeaks. Feeds mostly on insects.

Occurs in most regions in China except the northwest. A summer visitor, passage migrant and winter visitor in Beijing where it is rather common in every park.

中文名 **乌鸫**
学　名 *Turdus mandarinus*
英文名 Chinese Blackbird

雀形目
PASSERIFORMES
鸫科
Turdidae

乌鸫是一种中型鸣禽。雌雄相似。整体黑色，似小型乌鸦；眼圈黄色，虹膜深褐色，黄色喙较细长。雌鸟似雄鸟，但整体棕褐色。

乌鸫常栖息于平原地区的公园绿地、混交林等环境。常单独或成对活动，性大胆，不惧人。喜在地面觅食，主要以蚯蚓和昆虫为食。

乌鸫分布于我国除东北、西北、西藏以外的多数地区。北京城区各公园较常见，为留鸟。

保护级别 Protected-level	北京市重点保护 Beijing priority conservation	国家重点保护 State priority conservation	IUCN
	是（YES）	——	无危（LC）2018.8.9

A medium-sized songbird, sexes similar. Black overall like a smaller-sized crow; yellow eye-ring; dark brown iris; yellow and slender bill; female similar to male, but brownish overall.

Often inhabits environments such as green spaces in parks and mixed forests in plains. Usually solitary or in pairs. Bold and unafraid of people. Likes foraging on the ground; feeds mostly on earthworms and insects.

Occurs in most regions in China except the northeast, the northwest, and Xizang. A resident in Beijing where it is rather common in every park.

中文名 红尾斑鸫

学　名 *Turdus naumanni*

英文名 Naumann's Thrush

雀形目 PASSERIFORMES

鸫科 Turdidae

红尾斑鸫是一种中型鸣禽。雌雄相似。成鸟头顶、耳羽灰色，眉纹、下颏、颈侧红色；上喙及下喙尖端黑色，下喙基黄色；上体灰色；下体白色，具红色鳞状斑纹；尾偏红色。雌鸟似雄鸟，但整体色浅，少斑纹。

红尾斑鸫常栖息于平原地区及山区的公园绿地、针叶林、混交林、灌丛等环境。非繁殖期常成群活动，有时与斑鸫及其他鸫混群活动。主要以昆虫和植物果实为食。

红尾斑鸫分布于我国除海南以外的各地区。北京城区各公园及郊区较常见，为旅鸟和冬候鸟。

北京市重点保护 Beijing priority conservation	国家重点保护 State priority conservation	IUCN
—	—	无危（LC）2016.10.1

A medium-sized songbird, sexes similar. Adult has gray crown and auriculars; orange supercilium, chin, and side of neck; black tip on both maxilla and lower mandible; the latter has yellow base; gray upperparts; white underparts with orange scale-like markings; orange tail; female similar to male but duller overall; less markings.

Often inhabits environments such as green spaces in parks, coniferous forests, mixed forests, and shrublands in plains and mountainous areas. Form flocks in the non-breeding season, sometimes with Dusky Thrushes and other thrushes. Feeds mostly on insects and fruits.

Occurs in every region in China except Hainan. A passage migrant and winter visitor in Beijing where it is rather common in parks and suburban areas.

中文名 **斑鸫**
学　名 *Turdus eunomus*
英文名 Dusky Thrush

雀形目
PASSERIFORMES

鸫科
Turdidae

斑鸫是一种中型鸣禽。雌雄相似。成鸟头顶、耳羽、髭纹灰黑色，眉纹、下颏、喉部白色；上喙及下喙尖端黑色，下喙基黄色；背部灰褐色，两翼偏红色；下体白色，具灰黑色鳞状斑纹；尾偏灰黑色。雌鸟似雄鸟但整体色浅。

斑鸫常栖息于平原地区及山区的针叶林、混交林、灌丛等环境。非繁殖期常成群活动，有时与其他鸫混群活动。主要以昆虫和植物果实为食。

斑鸫分布于我国多数地区。北京城区各公园及郊区较常见，为旅鸟和冬候鸟。

北京市重点保护 Beijing priority conservation	国家重点保护 State priority conservation	IUCN
—	—	无危（LC）2016.10.1

成鸟 Adult

A medium-sized songbird, sexes similar. Adult has dark crown, auriculars, and malar stripe; white supercilium, chin, and throat; dark bill tip; base of lower mandible yellow; grayish brown back; rufous wings; white underparts with black scale-like markings; dark tail; female similar to male but duller overall.

Often inhabits environments such as coniferous forests, mixed forests, and shrublands in plains and mountainous areas. Form flocks in the non-breeding season; sometimes flocking with other thrushes. Feeds mostly on insects and fruits.

Occurs in most regions in China. A passage migrant and winter visitor in Beijing where it is rather common in parks and suburban areas.

中文名 **乌鹟**
学　名 *Muscicapa sibirica*
英文名 Dark-sided Flycatcher

雀形目
PASSERIFORMES
鹟科
Muscicapidae

乌鹟是一种小型鸣禽。雌雄相似。成鸟头部、背部灰黑色，喉部白色；喙较小，上喙黑色，下喙基黄色不明显；颈侧具一白色半颈环，两翼及尾羽黑色；胸腹部白色，具不规则的深灰色斑纹。

乌鹟常栖息于中低海拔山区、丘陵、平原地带的针叶林、混交林、林缘灌丛等环境。常单独活动，喜在林中层寻觅食物。主要以昆虫为食。

乌鹟分布于我国除西北以外的多数地区。北京城区各公园及郊区常见，多为旅鸟。

北京市重点保护 Beijing priority conservation	国家重点保护 State priority conservation	IUCN
—	—	无危（LC）2016.10.1

成鸟 Adult

A small songbird, sexes similar. Adult has dark head and back; white throat; rather small bill with dark maxilla and indistinct yellow base of lower mandible; white half-collar on neck; dark wings and tail; white chest and belly with dark and irregular markings.

Often inhabits environments such as coniferous forests, mixed forests, forest-edge shrublands in mountains, hills, and plains at low and middle elevations. Usually solitary. Likes foraging in forest midstory; feeds mostly on insects.

Occurs in most regions in China except the northwest. A passage migrant in Beijing where it is common in parks and suburban areas.

中文名 北灰鹟

学　名 *Muscicapa dauurica*

英文名 Asian Brown Flycatcher

雀形日 PASSERIFORMES

鹟科 Muscicapidae

成鸟 Adult

北灰鹟是一种小型鸣禽。雌雄相似。成鸟头部、背部灰黑色，喉部白色；喙较大，上喙黑色，下喙基黄色明显；颈侧半颈环不明显，两翼及尾羽黑色，胸腹部白色斑纹少。

北灰鹟常栖息于中低海拔山区、平原地带的混交林、次生林、林缘灌丛等环境。常单独活动，喜在林中层寻觅食物。主要以昆虫为食。

北灰鹟分布于我国除西北以外的多数地区。北京城区各公园及郊区常见，多为旅鸟。

保护级别 Protected-level	北京市重点保护 Beijing priority conservation	国家重点保护 State priority conservation	IUCN
	—	—	无危（LC）2016.10.1

A small songbird, sexes similar. Adult has grayish-black head and back; white throat; large bill with black maxilla and distinct yellow base of lower mandible; indistinct half-collar on neck; black wings and tail; little white spotting on chest and belly.

Often inhabits environments such as mixed forests, secondary forests, forest-edge shrublands in plains and mountainous areas in low and middle elevations. Usually solitary. Likes foraging in forest midstory; feeds mostly on insects.

Occurs in most regions in China except the northwest. A passage migrant in Beijing where it is common in parks and suburban areas.

084

中文名 **红胁蓝尾鸲**
学　名 *Tarsiger cyanurus*
英文名 Orange-flanked Bush-robin

雀形目
PASSERIFORMES
鹟科
Muscicapidae

红胁蓝尾鸲是一种小型鸣禽。雌雄差异大。雄鸟头顶、脸颊、背部、尾羽亮蓝色，两翼近黑色，眉纹白色，喙黑色，喉部、胸腹部白色，两胁橘红色。雌鸟整体灰褐色，两胁橘红色，尾羽蓝色。

红胁蓝尾鸲常栖息于低山及平原地带的针叶林、混交林、竹林、灌丛等环境。常单独活动。喜在地面觅食。主要以昆虫和植物果实为食。

红胁蓝尾鸲分布于我国除西南以外的多数地区。北京城区各公园及郊区常见，多为旅鸟和冬候鸟。

北京市重点保护 Beijing priority conservation	国家重点保护 State priority conservation	IUCN
是（YES）	—	无危（LC）2016.10.1

雌鸟 Female

A small songbird, sexes different. Male has bright blue crown, cheek, back, and tail; wings nearly black; white supercilium; black bill; white throat, chest, and belly; rufous flanks. Female grayish brown overall; orange flanks; blue tail.

Often inhabits environments such as coniferous forests, mixed forests, bamboo forests, and shrublands in low-elevation mountains and plains. Usually solitary. Likes foraging on the ground; feeds mostly on insects and fruits.

Occurs in most regions in China except the southwest. A passage migrant and winter visitor in Beijing where it is common in parks and suburban areas.

中文名 **红喉姬鹟**
学　名 *Ficedula albicilla*
英文名 Taiga Flycatcher

雀形目
PASSERIFORMES
鹟科
Muscicapidae

红喉姬鹟是一种小型鸣禽。雌雄差异大。雄鸟繁殖羽头顶灰褐色，脸颊灰色，喉部橙红色，上体灰褐色，胸部灰色，腹部灰褐色，尾羽近黑色，外侧尾羽基部白色。雌鸟似雄鸟，但喉部偏白色。

红喉姬鹟常栖息于低海拔山区、丘陵、平原地带的混交林、次生林、林缘灌丛等环境。常单独或成对活动，喜在林中层寻觅食物。主要以昆虫及植物果实为食。

红喉姬鹟分布于我国多数地区。北京城区各公园及郊区常见，多为旅鸟。

北京市重点保护 Beijing priority conservation	国家重点保护 State priority conservation	IUCN
—	—	无危（LC）2016.10.1

A small songbird, sexes different. Breeding male has grayish brown crown, gray cheek, and an orange throat; grayish brown upperparts; grayish chest and belly; tail almost black with the base of outer tail feather white. Female similar to male but with whitish throat.

Often inhabits environments such as mixed forests, secondary forests, forest-edge shrublands in low-elevation mountains, hills, and plains. Usually solitary or in pairs. Likes foraging in forest midstory; feeds mostly on insects and fruits.

Occurs in most regions in China. A passage migrant in Beijing where it is common in parks and suburban areas.

中文名 北红尾鸲

学　名 *Phoenicurus auroreus*

英文名 Daurian Redstart

雀形目
PASSERIFORMES

鹟科
Muscicapidae

雄鸟 Male

北红尾鸲是一种小型鸣禽。雌雄差异大。雄鸟头顶灰白色，喉部、背部黑色；两翼黑色，具一明显的白斑；胸腹部、腰部橘红色；中央尾羽黑色，两侧橘红色。雌鸟整体灰褐色，翼上具一白斑。

北红尾鸲常栖息于低山及平原地带的混交林、灌丛、村落及城市公园等环境。常单独或成对活动。主要以昆虫和植物果实为食。

北红尾鸲分布于我国除西北以外的多数地区。北京城区各公园及郊区常见，多为旅鸟、夏候鸟和冬候鸟。

保护级别 Protected-level	北京市重点保护 Beijing priority conservation	国家重点保护 State priority conservation	IUCN
	—	—	无危（LC）2016.10.1

A small songbird, sexes different. Male has whitish gray crown; black throat and back; black wings with a distinct white patch; orange chest, belly and rump; dark central rectrices and the remainder of tail orange-red. Female grayish brown overall with a white wing patch.

Often inhabits environments such as mixed forests, shrublands, villages, and urban parks in low-elevation mountains and plains. Usually solitary or in pairs. Feeds mostly on insects and fruits.

Occurs in most regions in China except the northwest. A passage migrant, summer visitor, and winter visitor in Beijing where it is common in parks and suburban areas.

087

中文名 **黑喉石䳭**

学　名 *Saxicola maurus*

英文名 Siberian Stonechat

雀形目
PASSERIFORMES

鹟科
Muscicapidae

黑喉石䳭是一种小型鸣禽。雌雄差异大。雄鸟头部黑色，颈侧白色；上体黑色，有一白色翼斑；下体沾橘红色，尾羽黑色。雌鸟整体偏灰褐色，翼上具一白斑。

黑喉石䳭常栖息于低山、丘陵、平原地带的灌丛、芦苇丛等环境。常单独或成对活动，喜在开阔地区的灌丛高处寻觅食物。主要以昆虫为食。

黑喉石䳭分布于我国东部的多数地区。北京城区各公园及郊区常见，多为旅鸟。

北京市重点保护 Beijing priority conservation	国家重点保护 State priority conservation	IUCN
——	——	无危（LC）2020.8.12

雌鸟 Female

A small songbird, sexes different. Male has black head; white side of the neck; black upperparts with a white wing patch; rusty-red underparts; black tail. Female grayish overall with a white wing patch.

Often inhabits environments such as shrublands and reedbeds in low-elevation mountains, hills, and plains. Usually solitary or in pairs. Likes foraging above shrubs in open areas; feeds mostly on insects.

Occurs in most regions in eastern China. A passage migrant in Beijing where it is common in parks and suburban areas.

088

中文名 **麻雀**
学　名 *Passer montanus*
英文名 Eurasian Tree Sparrow

雀形目
PASSERIFORMES
雀科
Passeridae

黑色斑块
Black spot

成鸟 Adult

麻雀是一种十分常见的小型鸣禽。雌雄相似。成鸟头顶棕色，白色脸颊具一黑色斑块，喉部黑色；上体棕褐色，具黑色斑纹；下体及尾羽棕色。

麻雀常栖息于城市及乡村地带有人类居住的各类环境中，为最典型的伴人鸟，非常适应在人类居住地生活。非繁殖期常成群活动。主要以昆虫及植物果实、种子等为食。

麻雀分布于我国多数地区。北京城区、郊区常见，为留鸟。

北京市重点保护 Beijing priority conservation	国家重点保护 State priority conservation	IUCN
——	——	无危（LC）2016.10.1

幼鸟 Juvenile

A very common small songbird, sexes similar. Adult has brown crown; white cheek with a black spot; black throat; rufous-brown upperparts have black streaking; brown underparts and tail.

Often inhabits populated urban and rural areas; a typical synanthropic bird; very adept at living in human-dominated environments. Form flocks in the non-breeding season. Feeds mostly on insects, fruits, and seeds.

Occurs in most regions in China. A resident in Beijing where it is common in urban and suburban areas.

089 中文名 树鹨

学　名 *Anthus hodgsoni*

英文名 Olive-backed Pipit

雀形目
PASSERIFORMES

鹡鸰科
Motacillidae

成鸟 Adult

树鹨是一种中型鸣禽。雌雄相似。成鸟头顶橄榄绿色，具黑色斑纹；眉纹皮黄色，耳羽有一白色斑块；上体橄榄绿色，具黑色斑纹，下体皮黄色及白色，具黑色纵纹；尾羽橄榄绿色。

树鹨常栖息于平原、丘陵、山区地带的针叶林、混交林、林间草地等环境。非繁殖期常成群活动，喜在地面觅食。主要以昆虫为食。

树鹨分布于我国多数地区。北京城区及郊区常见，多为旅鸟。

北京市重点保护 Beijing priority conservation	国家重点保护 State priority conservation	IUCN
—	—	无危（LC）2016.10.1

成鸟 Adult

A medium-sized songbird, sexes similar. Adult has olive crown with black streaking; yellowish supercilium; white spot on ear-coverts; olive green upperparts with black streaking; underparts yellowish white with black streaking; olive green tail.

Often inhabits environments such as coniferous forests, mixed forests, and woodland meadows in plains, hills, and mountainous areas. Form flocks in the non-breeding season. Likes foraging on the ground; feeds mostly on insects.

Occurs in most regions in China. A passage migrant in Beijing where it is common in urban and suburban areas.

090

中文名 **水鹨**
学　名 *Anthus spinoletta*
英文名 Water Pipit

雀形目
PASSERIFORMES

鹡鸰科
Motacillidae

水鹨是一种中型鸣禽。雌雄相似。繁殖羽头顶、耳羽灰色，眉纹沾粉色，上体灰色，下体沾粉色而少斑纹，跗跖黑色，尾羽灰色。非繁殖羽头顶浅灰色，眉纹白色，上体灰褐色，下体白色而多纵纹。

水鹨常栖息于平原、低山地带的河流、湖泊、沼泽等环境。常单独活动，喜在地面觅食。主要以昆虫为食。

水鹨分布于我国多数地区。北京城区及郊区可见，多为旅鸟和冬候鸟。

北京市重点保护 Beijing priority conservation	国家重点保护 State priority conservation	IUCN
——	——	无危（LC）2018.8.9

成鸟 Adult

A medium-sized songbird, sexes similar. Breeding plumage includes gray crown and auriculars; pinkish supercilium; gray upperparts; underparts tinged pink and with little streaking; black tarsi; gray tail feathers. Non-breeding plumage includes duller crown, white supercilium, grayish brown upperparts, and white underparts with streaking.

Often inhabits environments such as rivers, lakes, marshes in plains and mountainous areas at low elevations. Likes foraging on the ground; feeds mostly on insects.

Occurs in most regions in China. A passage migrant and winter visitor in Beijing where it can be seen in urban and suburban areas.

中文名 灰鹡鸰

学　名 *Motacilla cinerea*

英文名 Gray Wagtail

雀形目
PASSERIFORMES

鹡鸰科
Motacillidae

成鸟 Adult

灰鹡鸰是一种中型鸣禽。雌雄相似。成鸟头顶、脸颊深灰色，眉纹白色，上体深灰色，下体、腰部黄色，跗跖肉色，深灰色尾羽长。雄鸟喉部黑色，雌鸟喉部白色。

灰鹡鸰常栖息于山区及平原地带的溪流、沼泽、湿润草地等环境。常单独或成对活动，喜在地面觅食。主要以昆虫为食。

灰鹡鸰分布于我国多数地区。北京房山、怀柔、延庆等地较常见，为旅鸟、夏候鸟及冬候鸟。

北京市重点保护 Beijing priority conservation	国家重点保护 State priority conservation	IUCN
——	——	无危（LC）2016.10.1

A medium-sized songbird, sexes similar. Adult has dark gray crown and cheek; white supercilium; dark gray upperparts and yellow underparts and rump; pinkish tarsi; tail dark gray and elongated; throat black on male and white on female.

Often inhabits environments such as rivers, marshes, moist grasslands in plains and mountainous areas. Usually solitary or in pairs. Likes foraging on the ground; feeds mostly on insects.

Occurs in most regions in China. A passage migrant, summer visitor, and winter visitor in Beijing where it is common in places such as Fangshan, Huairou, and Yanqing.

中文名 **白鹡鸰**
学　名 *Motacilla alba*
英文名 White Wagtail

雀形目
PASSERIFORMES
鹡鸰科
Motacillidae

黑背亚种

白鹡鸰是一种中型鸣禽。雌雄相似。白鹡鸰在国内亚种较多，北京记录过至少 6 个亚种。成鸟整体由黑白两色或黑白灰三色构成。雌鸟似雄鸟，但整体色浅。

白鹡鸰常栖息于平原、丘陵、山区地带的沼泽、草地、农田等多种环境。常单独或成对活动，喜在地面觅食。主要以昆虫为食。

白鹡鸰分布于我国多数地区。北京城区及郊区常见，为旅鸟、夏候鸟及冬候鸟。

北京市重点保护 Beijing priority conservation	国家重点保护 State priority conservation	IUCN
——	——	无危（LC）2018.10.30

A medium-sized songbird, sexes similar. Many subspecies in China, with at least six in Beijing. Adult bird consists of black and white or black, white, and gray. Female similar to male, but duller overall.

Often inhabits environments such as plains, hills, mountainous marshes, grasslands, and agricultural fields. Usually solitary or in pairs. Likes foraging on the ground; feeds mostly on insects.

Occurs in most regions in China. A passage migrant, summer visitor, or winter visitor in Beijing where it is common in urban and suburban areas.

093 中文名 燕雀

学　名 *Fringilla montifringilla*

英文名 Brambling

雀形目
PASSERIFORMES

燕雀科
Fringillidae

雄鸟 Male

雄鸟过渡色 Non-breeding male

燕雀是一种小型鸣禽。雌雄差异大。雄鸟繁殖羽头部黑色，喙厚实，偏黄色；背部黄褐色，具黑色鳞状斑纹；两翼棕褐色，带黑白两色翼斑；胸部棕褐色，腹部白色。雌鸟和非繁殖期雄鸟整体体色较浅。

燕雀常栖息于平原、中低海拔山区的混交林、阔叶林、次生林、农田、公园等环境。非繁殖期常成群活动，有时可达千只。主要以昆虫及植物种子为食。

燕雀分布于我国多数地区。北京城区及郊区常见，多为旅鸟和冬候鸟。

北京市重点保护 Beijing priority conservation	国家重点保护 State priority conservation	IUCN
是（YES）	——	无危（LC）2016.10.1

雌鸟 Female

A small songbird, sexes different. Breeding male has black head and thick, yellowish bill; yellowish brown back with black scale-like markings; brown wings with black and white wing bars; orange chest and white belly. Female and non-breeding male duller overall.

Often inhabits environments such as mixed forests, broadleaf forests, secondary forests, agricultural fields, and parks in plains and mountainous areas at low elevations. Often form flocks in the non-breeding season, sometimes up to 1,000 individuals. Feeds mostly on insects and seeds.

Occurs in most regions in China. A passage migrant and winter visitor in Beijing where it is common in urban and suburban areas.

中文名 **黑尾蜡嘴雀**
学　名 *Eophona migratoria*
英文名 Chinese Grosbeak

雀形目
PASSERIFORMES
燕雀科
Fringillidae

雄鸟 Male

黑尾蜡嘴雀是一种中型鸣禽。雌雄差异大。雄鸟头部黑色面积较大；喙甚厚，黄色，喙尖黑色；枕部、胸部及腰部灰色，腹部皮黄色，背部褐色；两翼蓝黑色，翼尖白色；尾羽蓝黑色。两胁锈红色。雌鸟头部灰色。

常栖息于平原及低海拔山区的混交林、阔叶林、次生林、公园等环境。非繁殖期常成小群活动。主要以植物果实、种子为食。

黑尾蜡嘴雀分布于我国东部多数地区。北京城区及郊区可见，多为留鸟和旅鸟。

北京市重点保护 Beijing priority conservation	国家重点保护 State priority conservation	IUCN
是（YES）	——	无危（LC）2016.10.1

雌鸟 Female

A medium-sized songbird, sexes different. Male has large area of black on its head; thick, yellow bill with black tip; gray nape, chest, and rump; yellow belly, brown back; dark blue wings with white wing tips; dark blue tail; paler rusty-red flanks. Head of female gray.

Often inhabits environments such as mixed forests, broadleaf forests, secondary forests, and parks in plains and mountainous areas at low elevations. Form small flocks in the non-breeding season. Feeds mostly on fruits and seeds.

Occurs in most regions in eastern China. A resident and passage migrant in Beijing where it can be seen in urban and suburban areas.

中文名 **金翅雀**
学　名 *Chloris sinica*
英文名 Oriental Greenfinch

雀形目
PASSERIFORMES
燕雀科
Fringillidae

金翅雀是一种小型鸣禽。雌雄相似。雄鸟头部灰色，喙偏粉色，胸腹部黄褐色，背部棕褐色，两翼由金黄色、黑色、白色构成，尾羽黑色。雌鸟似雄鸟，但整体色浅。

金翅雀常栖息于平原及中低海拔山区的针叶林、混交林、次生林、公园等环境。非繁殖期常成小群活动。主要以植物果实、种子为食。

金翅雀分布于我国东部多数地区。北京城区及郊区常见，多为留鸟。

保护级别 Protected-level	北京市重点保护 Beijing priority conservation	国家重点保护 State priority conservation	IUCN
	是（YES）	——	无危（LC）2018.8.7

雌鸟 Female

A small songbird, sexes similar. Male has gray head, pinkish bill, yellowish brown underparts, and olive brown back; wings consist of golden yellow, black, and white; tail feathers black. Female similar to male, but duller overall.

Often inhabits environments such as coniferous forests, mixed forests, secondary forests, and urban parks. Form small flocks in the non-breeding season. Feeds mostly on fruits and seeds.

Occurs in most regions in eastern China. A resident in Beijing where it is common in urban and suburban areas.

096 中文名 三道眉草鹀

学　名 *Emberiza cioides*

英文名 Meadow Bunting

雀形目 PASSERIFORMES

鹀科 Emberizidae

三道眉草鹀是一种中型鸣禽。雌雄相似。雄鸟头顶、耳羽栗红色，眉纹、下颊纹、喉部白色；上体棕色，具深褐色斑纹；胸部棕红色，腹部、腰部及尾羽棕色。雌鸟似雄鸟，但整体色浅。

三道眉草鹀常栖息于丘陵及中低海拔多灌丛的山地环境。非繁殖期有时成小群活动。主要以植物种子、昆虫为食。

三道眉草鹀分布于我国除西南以外的多数地区。北京房山、延庆、怀柔等地较常见，为留鸟。

北京市重点保护 Beijing priority conservation	国家重点保护 State priority conservation	IUCN
是（YES）	——	无危（LC）2016.10.1

成鸟 Adult

A medium-sized songbird, sexes similar. Male has chestnut crown and auriculars; white supercilium, malar stripe, and throat; brown upperparts with dark streaking; brownish-red chest; brown belly, rump, and tail. Female similar to male, but duller overall.

Often inhabits environments such as hills and mountainous shrublands at low and middle elevations. Form small flocks in the non-breeding season. Feeds mostly on seeds and insects.

Occurs in most regions in China except the southwest. A resident in Beijing where it is common in places such as Fangshan, Yanqing, and Huairou.

中文名 **灰眉岩鹀**
学　名 *Emberiza godlewskii*
英文名 Godlewski's Bunting

雀形目
PASSERIFORMES
鹀科
Emberizidae

灰眉岩鹀是一种中型鸣禽。雌雄相似。雄鸟眉纹、脸颊、喉部蓝灰色，侧冠纹及贯眼纹棕色，上体棕色，多深褐色斑纹；下体及腰部棕褐色，无斑纹；尾羽近黑色。雌鸟似雄鸟，但头部色浅。

灰眉岩鹀常栖息于中低海拔的多石、灌丛、针叶林环境。非繁殖期有时成小群活动。主要以植物种子、昆虫为食。

灰眉岩鹀分布于我国北部的多数地区。北京房山、延庆、怀柔等地较常见，为留鸟。

北京市重点保护 Beijing priority conservation	国家重点保护 State priority conservation	IUCN
——	——	无危（LC）2018.8.9

成鸟 Adult

A medium-sized songbird, sexes similar. Male has blue-grey supercilium, cheek, and throat; brown lateral crown-stripe and eye-stripe; brown upperparts with dark streaking; brown underparts and rump have no streaking; tail nearly black. Female similar to male, but head duller.

Often inhabits environments such as rocky habitats, shrublands, and coniferous forests at low and middle elevations; form small flocks in the non-breeding season; feeds mostly on seeds and insects.

Occurs in most regions in northern China. A resident in Beijing where it is common in places such as Fangshan, Yanqing, and Huairou.

中文名 **黄喉鹀**

学　名 *Emberiza elegans*

英文名 Yellow-throated Bunting

雀形目 PASSERIFORMES

鹀科 Emberizidae

黄喉鹀是一种小型鸣禽。雌雄相似。雄鸟头顶具一黑色羽冠，黄色眉纹向后延伸至枕部，眼周至耳羽及胸部黑色，喉部黄色；上体棕色，多深褐色斑纹；下体白色；两胁具棕褐色纵纹。雌鸟似雄鸟，但头顶及耳羽褐色。

黄喉鹀常栖息于平原、丘陵及中低海拔的针叶林、混交林、灌丛、草地、果园等环境。非繁殖期常成小群活动。主要以植物种子、昆虫为食。

黄喉鹀分布于我国东部的多数地区。北京城区及延庆、密云等地较常见，为旅鸟、冬候鸟和夏候鸟。

北京市重点保护 Beijing priority conservation	国家重点保护 State priority conservation	IUCN
是（YES）	——	无危（LC）2018.8.9

雌鸟 Female

A small songbird, sexes similar. Male has black crest; yellow supercilium stretches to nape; black mask and chest patch; yellow throat; brown upperparts with dark streaking; white underparts with brown streaking in flanks. Female similar to male, but has grayish brown crest and auriculars.

Often inhabits environments such as plains and hills, as well as coniferous forests, mixed forests, shrublands, grasslands, and orchards at low and middle elevations. Form small flocks in the non-breeding season. Feeds mostly on seeds and insects.

Occurs in most regions in eastern China. A passage migrant, winter visitor, and summer visitor in Beijing where it is common in places such as Yanqing and Miyun.

中文名 **苇鹀**
学　名 *Emberiza pallasi*
英文名 Pallas's Reed Bunting

雀形目
PASSERIFORMES

鹀科
Emberizidae

苇鹀是一种小型鸣禽。雌雄相似。雄鸟繁殖羽头部、喉部黑色，白色髭纹向后延伸至枕部；上体棕色，具黑色斑纹；肩羽灰色；下体皮黄色；尾羽近黑色。雄鸟非繁殖羽及雌鸟侧冠纹及耳羽棕色，眉纹皮黄色；下体皮黄色，具深褐色纵纹。

苇鹀常栖息于平原地区的沼泽、河流、池塘地带的芦苇及高草地等环境。非繁殖期常成小群活动。主要以昆虫及植物种子为食。

苇鹀分布于我国东部多数地区及西北地区。北京城区及延庆、密云等地较常见，为旅鸟、冬候鸟。

北京市重点保护 Beijing priority conservation	国家重点保护 State priority conservation	IUCN
—	—	无危（LC）2018.8.9

雌鸟 Female

A small songbird, sexes similar. Breeding male has black hood and throat; white malar stripe stretches to nape; upperparts brown with black streaking and gray scapulars; sandy brown underparts; tail nearly black. Female and non-breeding male have brown lateral crown-stripe and auriculars, sandy brown supercilium and underparts with dark brown streaking.

Often inhabits environments such as reedbeds and tall grasslands in marshes, rivers, ponds. Form small flocks in the non-breeding season. Feeds mostly on insects and seeds.

Occurs in northwestern and most regions in eastern China. A passage migrant and winter visitor in Beijing where it is common in the urban areas and places such as Yanqing and Miyun.

100

中文名 **小鹀**
学　名 *Emberiza pusilla*
英文名 Little Bunting

雀形目
PASSERIFORMES
鹀科
Emberizidae

小鹀是一种小型鸣禽。雌雄相似。成鸟顶冠纹及耳羽栗色，耳羽外围及侧冠纹黑色，眉纹呈栗色（眉纹羽色于眼后变淡）；上体灰褐色，具红色、棕色斑纹；下体偏白色，具黑色纵纹；尾羽近黑色。

小鹀常栖息于平原、丘陵及中低海拔的阔叶林、次生林、灌丛、草地等环境。非繁殖期成群活动。主要以植物种子为食。

小鹀分布于我国多数地区。北京城区、郊区常见，为旅鸟。

北京市重点保护 Beijing priority conservation	国家重点保护 State priority conservation	IUCN
是（YES）	—	无危（LC）2016.10.1

A small songbird, sexes similar. Adult has chestnut median crown-stripe and auriculars; lower periphery of the auriculars and lateral crown-stripe black; supercilium chestnut in front and fading in the back; upperparts grayish brown with rufous and brown stripes; underparts whitish with black streaking; tail nearly black.

Often inhabits environments such as plains, hills, and broadleaf forests, secondary forests, shrublands, and grasslands at lower and middle elevations. Form flocks in the non-breeding season. Feeds mostly on seeds.

Occurs in most regions in China. A passage migrant in Beijing where it is common in both urban and suburban areas.

如何成为一名观鸟爱好者

近些年，观鸟这项老少皆宜的活动在我国逐渐兴起并快速发展。我国是世界鸟类资源最为丰富的国家之一，有约 1500 余种野生鸟类。北京的鸟类资源也非常丰富，有 515 种。

如何成为一名观鸟爱好者呢？说起观鸟，我们先聊聊什么是鸟？鸟类是体表被覆羽毛、有翼、恒温和卵生的高等脊椎动物。从生物学观点来看，旺盛的新陈代谢和飞行运动是其与众不同的进步性特征。旺盛的新陈代谢保证鸟类飞翔所需的高能量消耗，飞行运动能使鸟类迅速而安全地寻觅适宜栖息地或躲避天敌及恶劣自然条件的威胁，因此鸟类是陆生脊椎动物中分布最广、种类最多的一个类群。

简单地说，观鸟就是带着望远镜、图鉴和记录本，到自然环境中去观察、识别野生的鸟类。与其他自然观察相比，观鸟具备自身独特的魅力：首先，观鸟基本不受时间和地点限制，在我国北方的寒冬季节，昆虫很难被看到，多数植物也已凋零，而鸟类却还是相对常见。其次，对于“外貌协会”的朋友，观鸟可是您不错的选择！鸟类中，有些威猛霸气，有些可爱呆萌，有些羽色艳丽，总有一款适合您。另外，鸟类有着复杂的行为，观察鸟类的迁徙、觅食、求偶等各种行为，总是乐趣无穷。

望远镜是观鸟最重要的工具，一台适合您的望远镜可以让您快速掉入观鸟这个“大坑”。观鸟望远镜可分为两种：双筒望远镜和单筒望远镜。双筒望远镜轻巧，便于寻找鸟类，是观鸟的必备物品。单筒望远镜放大倍率更大，需要配合三脚架及云台使用，适合观赏距离较远的水鸟。如果您刚入门，单筒望远镜可以先不用考虑。

双筒望远镜分为两种，一种是保罗镜，一种是屋脊镜。保罗镜结构简单，较为便宜，但体积大，重量沉，观看时候的真实感较差。屋脊镜体积较小，更为轻便，观看时候更为舒服，但价格略高。综合来说，建议购买屋脊镜式望远镜。

观鸟望远镜绝对是个“大坑”，一支顶级的双筒望远镜动辄上万，有些顶级的单筒望远镜价格更是接近三万元。对于观鸟的初学者，建议去正规的望远镜店铺购买一支千元以内的望远镜就足够使用一段时间。

观鸟的第二个必备工具就是鸟类图鉴。鸟类图鉴之于观鸟，就相当于我们平时用字典查询生僻字。它可以告诉我们鸟的名字是什么，生活区域大概在哪里，生活习性是什么等诸多信息。随着国内观鸟事业的发展，鸟类图鉴也是琳琅满目，有些适合观鸟新人，有些适合观鸟达人。

观鸟记录本同样是我们观鸟的必备物品之一，也是部分观鸟爱好者最容易忽略的。一份完整的观鸟记录包括日期、地点、天气、温度、环境情况、观鸟起止时间、鸟的种类和数量等信息。这其中的每一个信息对于观鸟都可能有着很大的影响，随着观鸟记录的增多，我们可以更好地了解某些地区的鸟类变化规律。近年来，有一些观鸟记录中心可以完成数据实时传输，你可以每时每刻把自己的观鸟信息传递到网络上。不论是纸质记录还是电子记录，它们都是我们观鸟的重要财富。

随着数码相机的普及，拍鸟在我国逐渐风行起来，甚至拍鸟人的数量已远远地超过了观鸟人。同时，部分观鸟人士也会在观鸟之余进行鸟类摄影。传统的鸟类摄影以单反为主，其画质高，对焦快，连拍多，但对于观鸟人来说，某些长焦镜头过于沉重，高端机身和镜头价格较高。而长焦相机对于入门级的鸟友是不错的选择，亲民的价格、轻巧便携和较大光学变焦倍数是其优点，但这类相机对焦速度慢，成像质量较为一般。近年来，微单相机快速发展，它兼具了便携性和较高的成像质量，成为不少观鸟爱好者的钟爱神器。

如何成为一名观鸟爱好者，从下面几点说一说：

一、怎样发现鸟

在野外观鸟，找到鸟是第一步。但不同的鸟类可能栖息在不同的地理位置，偏好在不同的时间和不同的生境活动。那么我们要根据哪些信息找到更多的鸟呢?

1. 时间

俗话说“早起的鸟儿有虫吃”，这句话用在观鸟上也算得上名言警句了。的确，鸟类大多在清晨和傍晚活动，尤其是清晨，是很多鸟类的活动高峰，这是一天内最好的观鸟时间。当然所有的事情都不是绝对的，比如我们熟知的鸮，也就是民间俗称的猫头鹰，其中的多数种类就喜欢在夜晚活动。

2. 地点

地点从大范围来讲，可以理解为地理位置。首先你要知道，有些鸟种主要栖息于北方，有些鸟种主要栖息于南方。比如乌林鸮，这种大型的林鸮在我国仅分布于西北和东北地区。而另一种中大型的林鸮——褐林鸮，在我国则主要分布于长江以南地区。我国更多的鸟种属于候鸟，它们春季向北方迁徙去繁殖，秋季向南迁徙越冬。有些鸟喜欢栖息在高海拔地区，有些则喜欢栖息在低山或者平原地区。

3. 生境

不同鸟种的生活环境也有所区别。比如我国特有鸟种绿尾虹雉，主要分布在横断山脉的高海拔地区，喜欢在高山灌丛和高山草甸地带活动，如果在高山密林中寻觅这种美丽的雉类，遇见率可是不高。而在绿尾虹雉的分布区内还有一种名为红腹角雉的鸟类，它更喜欢在中高海拔的林下活动，如果去草甸寻找这种鸟，同样很难遇到。

二、观鸟观什么

1. 观鸟种

很多观鸟爱好者都把野外目击的鸟种数作为重要的追求目标，尤其在意一些罕见的种类。目击到个人新鸟种，是很多观鸟爱好者的最大乐趣。不过，过分追求新鸟种，而对看过的鸟不屑一顾，并不是一种好习惯。不仅失去了很多欣赏鸟类的机会，也偏离了观鸟的初衷，丧失了很多乐趣。

2. 观行为

鸟类的行为是非常值得去观察和思考的。如果鸟不动，那么到野外去观鸟与看标本、看照片就没有什么区别了。但鸟是会活动的，观鸟时，除了认识你所看到的鸟种，你还可以看它怎么飞，怎么跳，怎么觅食，怎么吸引配偶，怎么逃避敌害。

3. 观生态

很多观鸟达人之所以可以快速识别鸟类，其中一个重要的原因就是注重观察鸟类生境。鸟类的生存离不开生境，有些鸟会非常依赖某些特殊环境，比如长相似麻雀但却极度濒危的栗斑腹鹀，它的繁殖非常依赖山杏这种植物，如果你在其他环境寻找，很难看到它的身影。再比如我们在树林里看到一只圆圆胖胖的鹬飞过，那很有可能就是丘鹬，因为其他的相似种更喜欢在水边或草地活动。有鸟友说“观鸟就是观生境”，可见生境对于观鸟的重要性。

三、如何识别鸟种

作为观鸟新手，看到一只不认识的鸟，应该怎么观察才能识别呢？本书中《在野外如何识别鸟种》一文中有详细的介绍。

四、观鸟须知

（1）有些观鸟活动是在人迹较少的山区或者海边进行，要注意人身安全，不可单独行动，不可擅自进入陌生林区或滩涂等危险地带，不要接触鸟类粪便等排泄物。

（2）观鸟，是去野外观察、观赏野生鸟类，那些笼养鸟不被计入我们的观鸟记录当中。

（3）观鸟时如果遇到鸟类筑巢或育雏，要保持适当的观察距离，以免干扰鸟类繁殖，更不能采集鸟蛋。

（4）观鸟是一项安静的户外活动，活动中要保持安静，动作轻缓，不可高声叫喊或聊天，更不要扔石头惊吓鸟类。

（5）有人说观鸟不要穿红色、黄色等颜色鲜艳的服装，尽量选择与自然环境颜色近似的衣服。其实鸟的反应与您的衣服颜色关系并不是很大，与我们的行为相关性更大。如果在接近鸟时，扛着设备挺直身子迈着大步直奔鸟去，鸟在距离很远时就飞走了。尽可能放低身体，缓慢接近，这样做比选择衣服颜色要重要得多。

（6）拍摄野生鸟类时，尽量采用自然光，避免使用闪光灯，以免惊吓到鸟类或者干扰其正常行为。

（7）观鸟或拍鸟时，不可过分追逐野生鸟类。有些鸟可能因体能衰弱而暂时停栖在某一地区，此时，它们急需休息调养，过分的追逐行为，可能间接导致其死亡。

（8）爱护自然，不要随地吐痰、乱扔塑料制品，不要随意折树枝、采摘花朵。

观鸟是一项老少咸宜的活动。如果你有时间有兴趣，可以走入高山大川中去追寻难觅踪迹的罕见鸟种，也可以远渡重洋去异国他乡追寻新种。如果你没有那么多时间，也可以利用周末走入公园去观鸟，又或是在小区中观赏那些麻雀。观鸟是一件有趣的活动，不论何时何地，你总能看到那些美丽的精灵。观鸟又是一件严肃的活动，观察、识别、记录，都需要严谨负责的科学态度，一份份积累起来的观鸟记录，是很有意义的资料。观鸟收获最大的，不是看了多少鸟，走了多少地方，而是收获理性的思维和判断。

来吧，让我们拿起望远镜，走入户外，去观赏那些美丽的飞羽吧！

How to Become a Bird Watching Enthusiast

In recent years, bird watching, an activity suitable for all ages, has gradually emerged and developed rapidly in China. China is one of the countries with the richest bird resources in the world, with about 1,500 species of wild birds. Beijing is also very rich in bird resources, with 515 species.

How to become a bird-watching enthusiast? Speaking of bird watching, let's talk about what is a bird? Birds are high level oviparous vertebrates with wings, constant temperature, covered with feathers on the body surface. From the biological point of view, vigorous metabolism and the ability to fly are its distinctive progressive characteristics. Vigorous metabolism ensures the high energy consumption required by the birds to fly. Flying ability enables birds to find suitable habitats quickly and safely, or avoid the threat of natural enemies and harsh natural conditions. Therefore, birds are the most widely distributed and diverse taxa amongst terrestrial vertebrates.

In short, bird watching is to use telescopes, guidebooks and notebooks to observe, identify and record wild birds in the natural environment. Compared with other natural observations, bird watching has its own unique charm. Firstly, bird-watching is basically unrestricted by time and place. In the cold winter of northern China, insects are hard to be seen, most plants have withered, while birds are still relatively common. Secondly, for the friends of the "Appearance Association", bird watching is a good choice. Among the birds, some are powerful and domineering, some are cute, and some are colorful. There is always one suitable for you. In addition, birds have complex behaviors. It is always full of fun to observe their migration, foraging, courtship and other behaviors.

The most important tool for bird watching is a telescope. A suitable telescope can quickly attract you into this activity. First of all, bird-watching telescopes can be divided into two types, binoculars and monocular. Binoculars are light and easily used to find birds. It is a must-have item for bird watching. Monocular has greater magnification and need to be used with a tripod and head for viewing waterfowl at longer distances. If you're just getting started, monocular can be left off for now.

The binoculars can be divided into two types, the Paul prism and the roof prism. Paul prism is cheaper, simple in structure, large in size and heavy in weight. It has a poor sense of reality when being used watching birds. The roof prism is smaller, lighter, and more comfortable to view, however the price is slightly higher. In general, it is recommended to buy a roof prism binocular.

To buy a bird watching telescope will definitely spend you a lot of money. A pair of top binoculars often costs tens of thousands Yuan, while the price of a top monocular is close to 30,000 Yuan. For a junior birdwatcher, it is highly recommended to buy a pair of binoculars less than 1,000 Yuan in a regular telescope store, which can be used for a period of time.

The second essential tool for bird watching is the bird atlas. Bird illustrations are the equivalent of using a dictionary to look up obscure words. It can tell us about a bird's name, living area, habits, and so on. With the development of bird watching in China, bird atlases are also dazzling and plentiful. Some are suitable for the junior birdwatchers, some are suitable for the senior ones.

The notebook is also one of our must-have items for bird watching, and one that some birdwatchers are most likely to overlook. A complete bird watching record should include time, place, weather, temperature, environmental conditions, start time and end time, bird species and numbers, and some other information. Each of these may carry important information about the birds observed. As the bird watching records being accumulated, we can better understand the changing patterns of birds in certain areas. In recent years, there are a number of bird-watching recording centers that can accomplish real-time data transmission, so you can transmit your bird-watching information to the web every moment. Whether they are paper or electronic records, they are an important asset to our birding.

With the popularity of digital cameras, bird photography has gradually become popular in China, and even the number of bird photographers has far exceeded that of birdwatchers. Meanwhile, some birdwatchers also engage in bird photography in addition to bird-watching. Traditional bird photography is mainly based on DSLR, which means Digital Single Lens Reflex, with some significant advantages such as high picture quality, fast focus and more continuous shots. However, for birdwatchers, some telephoto lenses are too heavy, prices and prices for the matched cameras are all very high. But the telephoto camera is a good choice for the entry-level friends. The price is affordable, the weight is light, and the optical zoom range is large. But the focusing speed of this kind of camera is slow and the imaging quality is relatively average. In recent years, Mirrorless Interchangeable Lens Camera (MILC) developed rapidly. It has both portability and high imaging quality, making it a favorite magic weapon for many birdwatchers.

How to become into a bird watching enthusiast? Let's expound as follow:

• How do you spot a bird?

Finding a bird is the first step when it comes to looking at birds in the wild. But different birds may inhabit different geographic locations, preferring to be active at different times and in different habitats. So, what kind of information do we need to find more birds?

1.Time

As the saying goes, "The early bird catches the worm". This is even more suitable in bird watching. It’s true that most birds are much more active in the morning and the evening, especially early in the morning, which is the peak of activity for many birds, and this is the best time of the day for bird-watching. Of course all things are not absolute, such as different species of owls that we familiar with, most of them prefer to be active at night.

2. Locations

Locations can be interpreted as geographic location on a broad scale. The first thing you need to know is that some bird species inhabit mainly in the north and some in the south. For example, the Great Grey Owl, a large forest owl, is only distributed in the northwest and the northeast of China. Another kind of large and medium-sized forest owl, the Brown Wood-owl, is mainly distributed to the south of the Yangtze River in China. More of the bird species in China belong to migratory birds. They migrate to the north to breed in spring and to the south in autumn for the winter. Some birds like to live in high altitude areas, while others like to live in low mountains or plains.

3. Habitat

The living environment of different bird species is also different. For example, China's endemic bird Chinese Monal is mainly distributed at the high altitudes of the Hengduan Mountains. It likes alpine bushes and alpine meadows. The rate encounter with this beautiful pheasant in the dense alpine forest is not high. In the distribution area of Chinese Monal, there is another pheasant called Temminck's tragopan, which prefers stay under the forests at medium and high altitudes. If you go to the meadow to find this bird, it is also difficult to meet.

•What are you watching?

1. Identify the species

Many birdwatchers regard the number of bird species observed in the wild as an important goal, especially for some rare species. Witnessing a new bird species is the greatest pleasure for many bird watching enthusiasts. However, it is not a good habit to excessively pursue new bird species and ignore the birds you have ever seen. Not only lost a lot of opportunities to appreciate birds, but also deviated from the original intention of bird watching, lost a lot of fun.

2. Observe the behavior

The behavior of birds is well worth observing and thinking about. If the bird doesn't move in the wild, there is no difference from looking at bird specimens or photos. Fortunately all live birds in the wild are active. In addition to knowing the species of the bird you are observing, you can also watch how it flies, how it jumps, how it forages, how it attracts mates, and even how it escapes enemies.

3. Record the ecology

One of the important reasons many birdwatchers can quickly identify birds is to focus on bird habitats. Birds can't live if separate from its habitats. Some birds rely heavily on certain special environments, such as sparrow-like but extremely endangered Jankowski's Bunting. Its breeding depends heavily on the Siberian apricot. If you search in other environments, it is very hard to find it out. For another example, when we see a chubby sandpiper flying by in the woods, it is likely to be a Eurasian Woodcock, because the other similar species prefer to move by the water or the grass. That's why some birdwatchers said bird watching is sometimes habitat watching, which shows the importance of habitat for bird watching.

•How to identify bird species?

As a novice, how should I observe and recognize a bird I don't know? The article of *How to Identify a Bird in the Wild* in this book talks about it.

•Instructions for bird watching

(1) Some bird watching activities are carried out in mountainous areas or seaside with few people. Pay attention to your personal safety, do not act alone or enter dangerous areas such as unfamiliar forest or beaches without permission, and do not touch excrement such as bird droppings.

(2) Bird watching is to go outdoors to observe the wild birds. Caged birds

should not be counted.

(3) If you encounter birds nesting or raising young, you should keep an appropriate viewing distance to avoid interfering with bird breeding. What's more, collecting bird eggs is highly forbidden.

(4) Bird watching is a quiet outdoor activity. Keep quiet and slowdown when watching. Do not shout or chat loudly, let alone throw stones to scare the birds.

(5) Some people say we should not wear bright clothes such as red or yellow when birding and should try those with similar colors with the natural environment. In fact, the bird's reaction does not have much to do with the color of your clothes, but more relates to your behavior. If you carry your equipment, straighten your body, stride straightly when approaching the bird, it will fly away when you are still far away. Lower your body as much as possible and approach slowly, that is much more important than the color of your clothes.

(6) When shooting wild birds, try to use natural light while not flashlights, so that to avoid frightening birds or interfering with their normal behaviors.

(7) Do not chase wild birds too much when watching or shooting. Some birds may temporarily stay in a certain area due to physical weakness. They are in urgent need of rest and recuperation at this time. Excessive chasing behavior may indirectly lead to their death.

(8) Take care of the nature. Do not spit or throw plastic products. Do not fold branches or pick flowers.

Bird watching is an activity suitable for all ages. If you have time and are interested, you can walk into the mountains and by the rivers to find rare species of birds that are difficult to find, or you can fly across the oceans to the foreign countries to seek new species. If you don't have enough time, you can walk into the parks on weekends to watch birds or even watch those sparrows in the community. Bird watching is an interesting activity. You can always see those beautiful flying feather elves anytime anywhere. Bird watching is also a serious activity. Observation, identification and recording require rigorous and responsible scientific attitude. The accumulated bird watching records are very meaningful materials. The biggest gain through bird watching is not how many birds you have seen, how many places you have been, but to learn rational thinking and judgment.

Come on, let's pick up the binoculars, go outdoors, and watch those beautiful flying spirits covered with feather!

在野外如何识别鸟种

对于刚入门的鸟友来说，大家常常会面临一个困扰，那就是如何识别鸟类。鸟类的识别难度比较高，是因为有些鸟的相似度极高，像柳莺、树莺、鹟莺等，在野外较难识别；有些鸟比较相像，加之观察距离较远，不易看到识别要点，比如猛禽类；即使是同一种鸟，有些也会随着年龄、季节、性别而变化，比如鸻鹬类。下面对刚刚开始观鸟的鸟友简单讲解一下如何进行鸟类识别。

1. 大小

我们看到一只鸟，通常第一反应都是个体大小，这点是最为直接的判断，但在野外受观察距离的影响，需结合实际情况考量。刚开始观鸟时, 如果我们对大小不好界定, 可以建立几个参考鸟, 比如麻雀（约14cm）、白头鹎（约 18cm）、乌鸦（约 45cm）等。当你看到一只不认识的鸟类时，可以在笔记本上记录为：体长比麻雀略大等。

2. 体型

这是鸟类识别中非常重要的一点。前面说过，有些鸟类的羽色会随着年龄、性别、季节而变化，但它们的体型是相对不变的。比如猎隼与红隼比, 体型明显更加壮实。当然这需要一些野外观鸟经验的积累。

3. 喙

鸟类的喙有多种，如苍鹭、白鹭的喙长且直，戴胜、大杓鹬、棕颈钩嘴鹛的喙长而下弯，绿头鸭、白眼潜鸭的喙扁平，柳莺、树莺的喙尖细，红隼、苍鹰的喙呈钩状……再进一步，我们可以把喙描述得更加具体，比如北灰鹟的喙较乌鹟更大。

4. 翼

翼的形状也可以分为多种，圆形如纵纹腹小鸮，尖形如红隼，方形如普通鵟。同样，还可以进行对比，例如大鵟的翼较普通鵟的更长。

5. 尾

尾形更是多样，如平尾、圆尾、凸尾、凹尾、楔尾、叉尾等。但这里要注意两点：①描述尾羽要看其是合并时还是张开时；②考虑所观察到的鸟类是否处于换羽中。

6. 羽毛

很多人喜欢鸟类，是因为被它们多姿多彩的羽毛所吸引，而鸟类识别中，颜色也是识别要点之一。我们在记录鸟类的颜色时，要注意与鸟类的身体部位相对应。同时，有些鸟类的羽毛具有结构色，在不同光线下反射出的颜色不尽相同，这点要特别注意。

7. 姿态

当鸟在静止休息时，有些鸟类喜欢笔直而立，如猛禽、鵙；有些偏向于前倾，如杜鹃、歌鸲类；有些会仰头翘尾，如鹪鹩。

8. 飞行类型

有些鸟类在空中飞行时会采取波浪状飞行，如啄木鸟；有些则是盘旋，如普通鵟；有些喜欢垂直起落，如云雀；有些喜欢飞至空中转一周后落回原地，如多种鵙；还有些鸟类在迁徙时喜欢列队飞行，如鸿雁、普通鸬鹚、苍鹭等。

9. 鸣声

像柳莺、树莺类在野外不易区别，我们就可以凭借叫声来识别，尤其在春季，很多雀形目鸟类喜欢鸣唱，婉转的鸣声也是重要的识别要点。

10. 生境

有时候人们会问，有一只鸟从车头一闪而过，怎么有人就可以确定是什么鸟呢。这通常不仅需要一定的识别能力，还需要结合生境判断。例如在北京地区的林间惊飞一只沙锥状鸟类，有很大可能是丘鹬，这时候再加上颜色的判断，就可以比较准确地判断其身份。

最后我想说，野外鸟类识别是很有趣的。在识别过程中，我们不要怕出错，多交流，多学习，但确定鸟种时一定要谨慎。你的每一笔记录，无论对于自己观鸟之路，还是对于科学研究，都是非常重要的。

How to Identify Bird species in the Wild

Birders, especially the beginners, are often confused in how to identify and recognize a bird in the wild. It is often difficult to recognize a bird in the wild as some species of birds are quite similar. For instance, leaf warblers, tree warblers and flycatchers are hard to distinguish. Some species of birds look similar at first glance, and their characteristics may be vague from far distance. Even for the same species of bird, their appearance may change with aging, season and differ by the gender, such as shorebirds. Now we'd like to introduce some key point to beginners in recognizing a bird.

1. Size

The size of the bird is often the most recognizable feature. However, the size is often affected by distance, so more practical consideration is needed. If you are new in bird-watching and have less idea on assessing the size of a bird, you can choose some species of birds as reference. For example, sparrow (about 14cm), light-vented bulbul (about 18cm), crow(about45cm), etc, can be used as reference. When you see an unknown bird, you can record it in your notebook as slightly larger than sparrow in body length.

2. Shape

It is a very important characteristic in recognizing a bird, although it may take some practice. As discussed previously, the color of the bird may change in different age, season and gender. However, their body shape would remain unchanged. For instance, saker falcon is apparently stronger than common kestrel.

3. Beak

There are many types of bird beaks. Firstly, we can categorize bird beaks into several types. Grey heron and little egret have long and straight beak, hoopoe, curlew and scimitar babbler have long and down-curved beak, mallard and ferruginous duck have flattened beak, warbler has slender beak, and common kestrel and northern goshawk have hooked beak. Next, we can describe the beaks more specifically. For example, the mouth of Asian brown flycatcher is longer than dark-sided flycatcher.

4. Wing

The shape of wing can also be categorized into various types. The wing of little owl is round, common kestrel has sharp wing and eastern buzzard's wing is square. We can also make comparison between upland buzzard and eastern buzzard; the former has a longer wing.

5. Tail

There is a variety of shapes of the tails, including flat, round, convex, concave, wedge-shaped and fork shaped. There are two key points we should pay attention to, first, is it open or closed when we describe the tail shape, second, is it in molting time.

6. Feather

Many people like birds for their colorful feathers. It is also a main characteristic in recognizing them. We need to keep in mind that the description of feather color of a bird should be specific to the body part. At the same time, the feathers of some birds have structural color, which will be different on condition of reflecting sunshine from different angles; thus requires our special attention.

7. Posture

When the birds rest in quiet, some prefer to stand upright, such as raptors and flycatchers; some, for example, the cuckoos and robins, tend to lean forward. Birds like wrens would raise their head and tail while they perch.

8. Flying

Some birds, for example the woodpeckers, would fly in a waving manner, while some like to spiral in the air. For example, skylarks like to fly up and down vertically, and flycatchers often fly a circle in the air and land back to original spot. Some birds would fly in formation in migrating, such as the geese, cormorants and herons.

9. Singing and calling

Leaf warblers and bush warblers are not easily distinguished in the wild. We can recognize them by their singing or calling. Many passerine birds prefer singing and calling in spring, making their melodious song as an important characteristic.

10. Habitat

Many people are curious why someone could easily recognize a bird only by catching a glimpse of it flying over his head. It often requires the knowledge to recognize a bird, as well as the judgement of its habitat. For example, it might be a Eurasian woodcock if we startle a sandpiper shaped bird in a forest in Beijing. With its color, you can identify it more accurately.

Finally, it is very interesting in recognizing birds in the wild. We should not be afraid of making mistakes. You will learn more by sharing with others. Each of your record is important, not only for documenting your own progress, but also for scientific research.

野鸟救护常识

一、在野外遇到伤病鸟类的处理方法

1. 判断是否需要帮助

（1）如果有开放性外伤，或萎靡不振，没有行动能力，则需要我们的帮助。

（2）如果是春夏季节，幼鸟学飞时期，则需要进行判断：

如果全身绒毛，没有飞行能力，但可以蹦跳，可以放置在灌丛中或树枝上；如果全身绒毛，没有活动能力，需要我们帮助带走饲喂救护。

如果幼鸟羽毛长全，能短距离飞行，蹦跳自如，那就是正在学习飞行，不需要我们干预，可以在旁边观察，亲鸟稍后会将它带走。

（3）如果伤病鸟类旁边还有其他鸟类尸体，则可能是传染性疾病，不要接触，请维护好现场，及时通知动物防疫部门或救护机构。

2. 做好安全防护

在接触鸟类时，如果有条件，可以戴上手套；如果没有手套，在对鸟类处理完毕后，也请立即使用香皂或洗手液清洗消毒。

3. 安置鸟类

选择一个大小适中的纸箱安放鸟类，纸箱不应太大，以动物 2 倍体型大小为宜，在纸箱四周箱壁靠下方打一排透气孔，如有条件，可在箱子底部铺一条毛巾，供鸟类保暖或站立。如果是喜鹊类等尾巴较长的鸟类，还可以放置一段树枝。纸箱质地较软，不易使鸟类撞伤，且遮光，鸟类看不到外界环境，会觉得安全，减少其应激反应。

4. 喂水

一般鸟类在受到刺激后会大量脱水，如果鸟类体型较小，可以给它人工补液或喂水。左手（右手）虚握鸟类，控制住翅膀，使鸟类难以挣扎，但也不能太紧，避免使鸟类窒息。另外一只手用注射器或者毛巾沾水后滴向鸟类喙基（喙角），让鸟类自主吞咽，不可强行灌水，避免误入气管，导致二次受伤。

5. 喂食

一般情况下不要喂食，除非是刚出生不久的幼鸟，或者对该种鸟类十分熟悉。因为鸟类食性各不相同，而且大多数鸟类在被捕捉后会产生应激反应，拒绝进食。

6. 及时联系救护机构

最后，请及时联系专业救护机构：

北京市野生动物救护中心，010-89496118；

北京猛禽救护中心（仅限鹰隼鸮等鸟类），010-62205666。

二、如何救护幼鸟

1. 幼鸟出现

当夏天到来，很多幼鸟会出现在我们附近的草坪、绿地和灌丛，它们看起来是如此可爱，同时又是如此脆弱无助，没有丝毫抵抗外来危险的能力，您可能会心生怜爱，忍不住想把它们带回家照料。

但我们是否应该认真考虑一下这些幼鸟真正需要的是什么。

2. 幼鸟为什么离开巢

在可以飞行或自己捕食之前，多数幼鸟会离开它们的巢，为什么呢？因为随着幼鸟的生长，巢对于它们而言是一个越来越危险的地方。当它们的父母带着食物回来，这些叽叽喳喳的小家伙伸着脑袋索食的时候，巢变得不再具有隐蔽性，在松鼠、猛禽，甚至流浪猫的眼中，躲在巢中嗷嗷待哺的幼鸟是一道充满诱惑的美味大餐。

这时幼鸟选择离开巢，来到地面或者其他地方是很正常的现象。从此以后他们不再是嗷嗷待哺的雏鸟，因为它们已经长齐了羽毛，成为学飞的幼鸟。

这时父母会将它们的孩子带到附近的灌丛或者树林等隐蔽场所，然后继续照顾他们，这样持续几个星期后，幼鸟就能够自己独立寻找食物了。

3. 不要收养

鸟爸爸和鸟妈妈是非常关心幼鸟的，并且他们比我们更清楚什么才是幼鸟需要的，所以如果看到未受伤的幼鸟请不要收养，它们不是孤儿。

4. 您的帮助

不过在一些特殊的情况下，您还是可以提供一些帮助的，尤其是您发现幼鸟附近有流浪猫的踪迹。这时您可以将幼鸟转移到一个相对隐蔽的地点。

把它捡起来放在手中或毛巾里，一定要握稳，以防止幼鸟因挣扎而受伤，将它放进附近的密灌丛或矮树上，它自己便会找到安全的地方。

不必苛求把幼鸟放回巢内，因为无论把它放在什么位置，即使你将它们放到隔壁的邻居家或者街道对面，亲鸟也能通过它们的哭叫声找到它们，并且给它们带来食物。

5. 团聚

如果您的孩子带了幼鸟回家，您可以将它放回发现地，并且越快越好，即使它在您家里已经待了几个小时，或者一天，没有关系，只要您能将它送回到其父母身边，对于幼鸟来说就是最好的结果了。

您不用担心幼鸟的父母会拒绝它们，鸟类的嗅觉比我们人类要迟钝很多，它们不会嗅到您留在幼鸟身上的气味。它们十分期待孩子的回归。

如果您愿意的话，可以躲在一旁观察，在1至2个小时之内，幼鸟的父母就会找到它们的孩子，一家团聚了。

6. 美好的意愿

救助是件好事，但接下来的事就让大自然去做吧。

How to Rescue a Bird in Wild

• When you meet a wounded or sick bird in wild, you could take steps as follows

1. You should decide firstly whether the rescue is necessary

(1) If the bird has open wound, seems exhausted, or unable to fly or jump, it likely needs your help immediately.

In late spring or the early summer, when a young bird is practicing its flying skills, you need to make the following evaluation:

(2) If the young bird is still in its down feather, is incapable of flying, but it is already able to walk or jump, then you could put it in bushes or on the branch of a tree. Only those that can't move themselves need to be taken away and fed by the rescuer.

If the young bird has developed whole feathers, and has been able to fly short distance, jump and walk, it's better not to disturb it, you could stand by and observe it for a while, its parent may appear and attend to it.

(3) If an injured bird is accompanied with other dead birds, there may be infectious disease, please don't touch them without proper protection. You will need to inform the animal protection services and rescue agencies as soon as possible.

2. Proper protection measures

You should youar gloves while touching birds in the wild. If gloves are not available, you could wash your hands with soap and water or sanitize your hands after touching the birds with your hands.

3. Taking care of the rescued bird

The rescued bird should be placed in paper box that should be neither too big nor too small, approximately twice the size of bird. A row of ventilation hole could be made on the four sides of box, and a toyoul at the bottom of the box can keep the birds warming and standing. If it is the birds with a long tail such as magpie, you can place a branch in the box. Paper box is soft and can keep the birds away from injuring. The birds in the box can't see the outside, therefore they will feel safer and have less stress reaction.

4. Giving water

Birds will dehydrate a lot after stress. Small sized birds can be fed with water or with saline. You can hold the bird with one hand to control its wings in order not to let it struggle, without being too tight to suffocate it. Then you feed water with a syringe or yout toyoul with the other hand onto the base of the bird bill (corner of bird beak). You should keep it in your mind that you couldn't force the birds to drink, for it may cause injury if water enter the lungs.

5. Feeding

Generally speaking, there is no need to feed the bird except for newborns, or you are quite familiar with that species of bird, as different bird has different kind of food preference or eating behavior. Most birds would refuse to eat anything after being captured because of stress reaction.

6. Contacting rescue agency

Please contact your professional rescue agencies:

Beijing wildlife rescue center, 010-89496118

Beijing raptor rescue center (limited to birds such as hawk, falcon, owl)

010-622205666

• Rescue of young birds

1. Meeting the juvenile

Lots of young birds will appear in the lawns, green-belts or bushes around us during the summer time, they are so lovely but vulnerable at the same time. The temptation to bring them home and feed them in person is irresistible.

Should you think twice about what those young birds really need?

2. Why would young birds leave their nest

Why would most of young birds leave their nest before they can fly or get food on their own? In fact, living in the nest may become more and more dangerous as they grow up. When their parents come back with food, young birds may fight for food. The noises would make their nest more exposed to predators such as squirrels, raptors, even straying cats. Those little birds hiding in the nest may become delicious meal.

Therefore, it is quite common for young birds to leave their nests and stay on the ground or other spots. As they have developed full of feathers all over the body and have begun flying, they are becoming more independent and no longer completely live off their parent.

At that time, the parents would bring these young birds to nearby bushes or forests, and continue to take care of young birds for a few youeks until young birds can live on themself independently.

3. Don't adopt young bird

Parents are extremely concerned about their children. The same for the birds, and they know better than us what their children want. If you encounter uninjured young birds, please leave them alone, as they are not orphans.

4. Your help

Your help is needed in certain conditions, especially when stray cats or other dangerous animals may wonder around the young bird. In that case, you can take the bird to a relatively safe spot.

You can put the bird in hand or on a toyoul, hold it firmly, prevent it being hurt while the bird may struggle in your hand. You may place it in bushes or trees nearby, the bird could find a better place to hide.

It is not necessary to return the bird to its nest as his parents would always find it by following its cry. No matter where you place it, even if it is your neighbor's home or on the opposite side of the street, its parents can always track it and bring some food.

5. Reunion

If you happen to have brought a young bird home, you should bring it back to where it was found as soon as possible. It doesn't matter it is a few hours or a day he has been separated with his parent, as long as he could be sent back.

There is no need to worry about that the bird might be refused by its parent, as birds' smelling capability is less sensitive than humans. The parents cannot trace the odor you left on the bird and they are much willing to embrace their missing babies.

If possible, you can spend a few minutes to see what would be going on near around. Most likely, the parent would find their child in one or two hours, then the family get their reunion.

6. Good will

It is much meaningful to do rescuing works for birds or other animals in wild. Hoyouver, after the rescue, you should let the nature take its course.

野生鸟类的招引和保护

野生鸟类的招引和保护，最重要的是恢复和保护鸟类的栖息地。

无论是在山区，还是在平原，不同的鸟种需要不同类型的栖息环境。森林、草原、湿地、荒地、农田等都是鸟类不可或缺的栖息地。生态林的建设，优先考虑的是本土植物，无论是乔木、灌木、草被都应在人为监测和干预下尽快形成符合生物多样性的良好植物群落。在植被的发育过程中，各种无脊椎动物、脊椎动物都会逐渐回到它们的栖息地。湿地的管理和建设要注重水质保护及水生动植物的保护。严格控制农药的施用和化学污染，严格执行野生动植物保护法，都是加强生态文明建设的当务之急。

城市是人类活动最为集中的地方。随着生物多样性恢复与保护观念的普及，如何在宜居与环境友好方面对城市进行科学规划与建设，成为摆在我们面前的难题。我们提出关心城市鸟类，不只是吸引些鸟类在城市安家，而是希望通过对环境敏感的鸟类招引与保护，带动更多的本地物种回归城市，与我们共同生活，和我们共享这片土地，实现物种的丰富与多样，使城市更具生命活力，真正实现生态城市的愿景。

鸟类消失的主要原因是栖息地的改变和丧失，在北京城市中繁星一样分布的公园、绿地不仅能供人们休闲娱乐，还应能满足鸟类和更多物种的生存需求，成为各种动物的生存港湾，这就需要实现生态园林的布局和建设。

城市绿地的面积与质量直接影响生物的丰富程度。人们越来越认识到生态城市与生态园林的重要价值，为了吸引鸟类、昆虫及各种小动物，要为它们保留和种植更多的乡土植物，包括当地的乔木、灌木、草被，为鸟类和小动物们提供丰富的食物以及隐蔽、栖息、繁殖的环境条件，甚至要专门在公园和城市绿地中配置“本杰士堆”。

悬挂人工巢箱是一种有效的招引鸟类的技术手段，可以为那些找不到洞穴且赖以洞穴繁殖的鸟类提供巢址，弥补幼林区域及城市缺少大树天然洞穴的不足，也是目前人们普遍使用的人工招引方法之一。

人工巢箱的配置要有针对性，如为山雀、麻雀、姬鹟、椋鸟、雨燕、角鸮（猫头鹰）、鸳鸯等设计的巢箱，其规格及洞口的位置与大小都有差别。人工巢箱的颜色要融入周围的环境。制作巢箱的材料要比较讲究，要考虑经受风吹、日晒、雨淋等问题。木质材料和瓦罐制作的人工巢箱比较适宜，前者多需要涂刷棕、墨绿、黑色防水漆，后者保留瓦罐烧制的自然颜色即可。这两种材料透气性好，隔温性好，适宜鸟类筑巢繁殖。巢箱上部要设有遮挡雨水的盖板，巢箱底部要留有排水孔，以防雨水淹巢。雨燕的巢箱还要设有方便进出抓握的台子。为了方便对人工巢箱进行科学管理，巢箱应该便于开闭，并定期对巢箱进行检查与清理。如需要和条件允许，可以为人工巢箱安置低照度的视频监控与采集设备，在不干扰鸟类生活的情况下，采集它们的一些活动数据，为保护鸟类做科学资料的积累。鸟类对人工巢箱需要有一个熟悉和接受的过程，对于北京地区而言，悬挂人工巢箱应该在秋、冬季。

设置鸟食台（饲鸟器）是帮助城市鸟类度过严冬和食物匮乏时期的有效措施。各式各样的鸟食台是许多国家和地区的风靡之物，它们被悬挂在自家花园、公共绿地和城市公园，以吸引大量不同的鸟类前来“用餐”，特别是在大雪和严酷的季节，对鸟类的帮助特别大，减少和避免了鸟类的大量死亡。

饮水器和水盘的作用也不可小瞧，这些看似简单的装置，可以为生活在附近以及匆匆赶路，迁徙的鸟类提供必要的饮水。饮水器以使用清洁且流动的水为宜，既保证了野鸟的安全饮水，又避免了鸟类传染病的传播。

悬挂人工巢箱、配置喂鸟器和饮水器等是野生鸟类招引和保护的一些有效人工措施，但其仍然是保护鸟类的辅助手段。加强生态环境的建设，控制农药的施用，加快鸟类栖息地的恢复，改善生物多样性的状况，才能使鸟类、昆虫及其他动物都回到山地森林和我们居住的城市里。

Attraction and Protection of Wild Birds

The restoration and protection for the habitats of wild birds are the most important part in attracting and protecting works.

Different species of bird require a variety of habitats, whether it is in mountainous areas or plains. Forests, grasslands, wetlands, wastelands and farmlands are all essential habitats for birds. Priority should be given to local plants in the construction of ecological forests. A better biodiversity consists of local trees, shrubs and grasses that can be developed more quickly under our supervision. Various insects, other invertebrates and vertebrates will gradually return to their habitats as the local vegetation grows. We should pay more attention to water quality, aquatic plant and animals in the construction of wetland. We should strictly limit the use of pesticide and chemical pollution, firmly abide by wildlife protection law, all these are urgent tasks for further strengthening the construction of ecological civilization.

Cities are the places where human activities are most concentrated, and with the popularization of the concept of biodiversity restoration and conservation, how to scientifically plan and build cities in terms of livability and environmental friendliness has become a difficult problem in front of us. We come up with the idea of caring about urban birds, not only to attract urban birds to resettle in the city, but also for the recovery of more local plants, insects, fishes, amphibians, reptiles and mammals in our city, eventually to achieve the goal of their living and sharing with us. Only in this way can our city become an ecological city full of vitality.

The disappearance of birds in city is mainly caused by the change and loss of their habitats. Apart from providing people with leisure and entertainment, the parks and green lands scattered in city should be a haven and meet the need for survival of bird and other species, that demand the project and construction of ecological gardens.

The area and quality of green-land determine the biological richness. We are increasingly aware of the value of ecological city and gardens. More local plants including trees, shrubs and grasses should be planted and well protected in order to attract birds, insects and various animals. All of these offer foods and places for bird hiding, resting and breeding. Sometimes, even a "Benjeshecken" in park or green-land in city are needed.

Hanging man-made nests is an effective way to attract birds. These nests can help birds who lack tree-hole for breeding due to the shortage of forests and large trees in the city. This method is widely used in attracting bird at present.

Man-made nest should be specifically designed and made. The shape, size and location of entry are specific for different species of birds, such as tit, sparrow, flycatcher, starling, swift, owl and mandarin duck. Its color should be in line with the surroundings. The materials used should be more meticulous and resistible to wind, sunshine and rain. Wooden material and terracotta are commonly used. The former needs being painted in brown, dark-green or black water-proof paint and the latter could be kept in its original color. The two kinds of materials are good at ventilation and insulation, so they are suitable for birds to produce and breed. The upper part of the nest should have a cover to prevent rainwater to leak, and several holes should be dug in the bottom to drain water. The nest for swift needs to add a special platform for it to grasp easily. The man-made nest should be opened and closed conveniently, as we need to inspect and clean the nest regularly for maintenance. If needed, we can equip the nest with lower brightness monitor and collect data of their activities for scientific study, as long as they are not disturbing the birds. The birds themselves will gradually adapted to man-made nests. In Beijing, the best time to install the nest is autumn and winter.

Bird feeder is efficacious for birds to survive the harsh winter when food is scarce. It is common to see various kinds of feeders in family yard, public green-land and urban parks in many countries and regions. Feeder attracts a large number of birds to come for food. It contributes greatly in saving lives of birds in heavy snow days and harsh seasons.

Water fountain and water tray also play an important role. They provide necessary drinking water to birds living around and migrating, although these devices are seemingly simple. It is better to supply clean and flowing water. This will assure the safety of drinking for bird, as well as avoid the spreading of infectious disease.

Man-made nest, feeder, water fountain and other methods are effective in bird protection, but they are still auxiliary. Only by strengthening the construction of ecological environment, controlling the application of pesticide, speeding up the restoration of bird habitat, improving the status of biodiversity, can more and more birds, insects and other animals return to mountains, forests and our cities.

鸟名中的生僻字 Glossary

（按第一字笔画数排序）

六至十画

鸠 jiū(究）

鸢 yuān（冤）

鸤 shī（师）

鸬 yán（炎）

鸫 dōng（冬）

鸱鸮 chī xiāo（吃消）

鸬鹚 lú cí（炉磁）

鸲 qú（渠）

隼 sǔn（损）

十一至十五画

鸻 héng（恒）

鵟 kuáng（狂）

椋 liáng（凉）

鹡 jí（吉）

鹀 wú（吴）

鹎 bēi（杯）

鹊 què（却）

鹛 méi （眉）

鹡鸰 jí líng （急灵）

鹟 wēng （翁）

鹞 yào （药）

十六至二十画

鹨 liù（六）

鹪鹩 jiāo liáo（交辽）

鹬 yù（玉）

鹭 lù（路）

䴙䴘 pì tī（辟梯）

主要参考文献 Key References

北京市园林绿化局，北京市农业农村局．北京市重点保护野生动物名录 [EB/OL].（2022-12-31）[2023-04-09]https://yllhj.beijing.gov.cn/zwgk/fgwj/gfxwj/202301/t20230110_2894842.shtml

国家林业和草原局，农业农村部．国家重点保护野生动物名录 [EB/OL].（2021-02-25）[2023-04-08]http://www.forestry.gov.cn/main/3957/20210205/153020834857061.html

刘阳，2021. 中国鸟类观察手册 [M]. 长沙：湖南科学技术出版社．

张荣祖，2011. 中国动物地理 [M]. 北京：科学出版社．

赵欣如，2018. 中国鸟类图鉴 [M]. 北京：商务印书馆．

赵欣如，朱雷，黄瀚晨，等，2022. 北京鸟类图谱 [M]. 北京：中国林业出版社．

郑光美，2023. 中国鸟类分类与分布名录 [M].4 版．北京：科学出版社．